Matters
Mathematical

Matters Mathematical

I. N. Herstein
I. Kaplansky
University of Chicago

HARPER & ROW, PUBLISHERS
New York Evanston San Francisco London

Sponsoring Editor: George J. Telecki
Project Editor: Karen A. Judd
Designer: Howard S. Leiderman
Production Supervisor: Will Jomarrón

Matters Mathematical

Library of Congress Cataloging in Publication Data
Herstein, I N
    Matters mathematical.
    Includes bibliographies.
    1. Mathematics—1961–   I. Kaplansky, Irving,
1917–   joint author. II. Title.
QA39.2.H48   510   74–3755
ISBN 0–06–042803–1

Contents

Preface

This book is based on notes prepared for a course at the University of Chicago. The course was intended for nonmajors whose mathematical training was somewhat limited.

Our aim is to cover a selection of topics that give something of the flavor of modern mathematics. Furthermore, we try to carry each topic far enough to prove something substantial. Mastery of the material requires nothing beyond the algebra and geometry normally covered in high school.

In transforming the notes into a book, we expanded the material considerably. As a result we feel that the book may be suitable for a wider audience. For instance, it could be used in courses designed for students who intend to teach mathematics. Several of the topics touch directly on matters that the prospective teacher might be teaching his own students: geometry, probability and game theory, and portions of the number theory. Another possible use of the book is in a course for mathematics majors as a precursor to more specialized courses. Permutations, groups, and infinite sets could serve as their introduction to abstract mathematics.

We want the reader to see mathematics as a living subject in which new results are constantly being obtained. Whenever possible, we mention the most recent advances in the area under discussion.

Number theory has fascinated mankind for centuries. The subject abounds in open questions that are easily stated and easily understood. We open our long Chapter 2 with a preliminary informal section that surveys some of these challenges. This should motivate the more careful treatment that follows.

In much of modern mathematics one studies abstract systems. This is done not for the sake of generality alone, but because it has been found to be effective in solving classical problems. The abstractions evolved from special situations. A notable example is the transition from permutations to groups. Chapters 3 and 4 illustrate this: a "concrete" special case as a prelude to an abstraction. Permutations are easily grasped. They lend themselves to a variety of charming experiments; and nontrivial theorems and challenging open questions are not at all lacking. In this way the ground is prepared for the general concept of a group.

In Chapters 5 and 6 we treat two topics that are nearly self-contained: finite geometries and games. This is very much twentieth-century mathematics, and yet a minimum of technique is needed as a prerequisite. The high spot of Chapter 5 is the Bruck-Ryser theorem; that of Chapter 6 is the theory of two-by-two games.

Many students are introduced to set theory in grade school or high school. Chapter 1 reviews this material for such students and establishes the language we use throughout the book. We felt that it would be interesting to push beyond this modest level into the Cantor theory of infinite sets. People find infinity mysterious. The little that Chapter 7 contains on the cardinal equivalence of infinite sets sheds some light on these mysteries and introduces the reader to a remarkable chapter of current mathematics.

We try to avoid manipulation and technical details as much as possible, keeping the central ideas in the foreground. This cannot always be done. To get to the crux of a substantial piece of mathematics requires hard work and technique. But we believe that the book is within reach of people with the background we mentioned at the outset.

We owe a considerable debt to our colleague Melvin Rothenberg, who collaborated in the initial stages of the project, and to numerous students who made valuable comments and suggestions. In connection with the Josephus permutation (Chapter 3) we are grateful to Victor Akylas, Michael Bix, Alex Kaplansky, and Lucy Kaplansky for helpful hand computations, and to Raphael Finkel for preparing and running a program on the University of Chicago computer.

I.N.H.
I.K.

I'm very well acquainted too with matters mathematical,
I understand equations, both the simple and quadratical,
About binomial theorem I'm teeming with a lot o' news—
With many cheerful facts about the square of the hypotenuse.

Sung by the model of a modern Major-General
in Gilbert and Sullivan's *The Pirates of Penzance*

Chapter 1

Sets and Functions

In this first chapter, two subjects will be discussed: *sets* and *functions*. These concepts serve as a basic part of the language of today's mathematics. In later chapters we shall often use fundamental facts about sets and functions, and when this happens there will usually be no reference made to Chapter 1.

The main thing the reader should get from Chapter 1 is familiarity with the ideas, and once he has sufficient familiarity the mathematical difficulties should largely disappear. Something of this sort usually happens when the basic part of any subject gets developed.

But it should be added that considerable mathematical theories do exist concerning sets and functions. *Set theory* is a doctrine that has been developed to a high level of subtlety and sophistication. The first steps in this type of set theory are taken in Chapter 7. A close study of functions, as independent objects, began in the 1940s and grew into the branch of mathematics called *category theory*.

The first of the three sections in this chapter is devoted to basic facts about sets, the second gives applications to combinatorial problems (such as counting), and the last treats functions.

1

1. Sets

One way to explain what we mean by a "set" is to offer many equivalent words, such as "class," "collection," "aggregate," "flock," and "bevy." Perhaps we might add exotic ones, such as "pride." The resources of other languages might be called upon: "ensemble" in French, "Menge" in German, "kvutza" in Hebrew.

But perhaps it is best to learn from examples. Our examples will be mainly from mathematics (numbers and geometrical objects) but will also, on occasion, refer to the "outside" world.

Think of the numbers 2, 4, and 7. Imagine them gathered in a package for our inspection. If there are a large number of things we plan to do with the package, we might find it handy to label it, perhaps using a letter such as A. Then we can say that the set A consists of the numbers 2, 4, and 7. To put this in a more condensed way, we adopt the notation of enclosing 2, 4, and 7 in braces, and write $A = \{2, 4, 7\}$.

We speak of 2, 4, and 7 as being the *members* of A, or, alternatively, we say that they *belong* to A. Thus, 4 is a member of A; 4 belongs to A. It is a little too clumsy and long-winded to keep using these phrases; we wish to present the information in a more condensed way. There has been a nearly universal agreement to use ε (epsilon), the fifth letter of the Greek alphabet, for this purpose. We can now write, briefly, $4 \varepsilon A$ (the suggested reading is "four epsilon A"). Likewise we have $2 \varepsilon A$ and $7 \varepsilon A$. For the number 3 the statement $3 \varepsilon A$ does not hold. As a notation for this, we put a slanting line through the epsilon: $3 \notin A$. In the same way, $11 \notin A$, $0 \notin A$, and $-5 \notin A$.

For the set A that we have been discussing, it is extremely simple to list all of its members, and we are able to exhibit A by writing $A = \{2, 4, 7\}$, making it quite clear what set we are talking about. For other sets, things may not be so simple. For instance, the list of members, though explicit, might be extremely long. Or, we might be given a description of a set in such a way that the set is perfectly well determined, but it would require a lot of work to produce a list of members. Here are two nonmathematical examples:

> The set of all men now living who were born in Philadelphia in 1941.
> The set of all major league pitchers who won 20 or more games in 1965.

In the first case, the set is well defined but it would be quite a job to get an accurate list of its members. In the second case, listing the members is easy—get the right record book.

We return to sets that are composed of numbers. In the following example, a set is given by a description, but it is very easy to replace the description by a list of members. Let the set B be determined by saying that *B* is the set of all even numbers between 1 and 13. As an exercise at this point, we ask the reader to list all the numbers *x* such that *x* ε *B*. How many are there? Fill in the blank between the braces in *B* = {   }. Write several *y*'s such that *y* ∉ *B*.

For a similar but much bigger example, let *C* be the set of odd numbers between 2 and 1002. It would be useless and boring to write all the members of *C*. Instead we can exhibit enough of them so that the pattern becomes clear, putting dots in place of the rest:

$$C = \{3, 5, 7, 9, 11, \ldots, 997, 999, 1001\}$$

As a little exercise in arithmetic, we invite the reader to find how many numbers there are in *C*.

Next, let *D* be the set of *all* even positive numbers. When we try to make a list of the members of *D* we face a new problem: The list goes on forever. We can write

$$D = \{2, 4, 6, 8, 10, 12, \ldots\}$$

but now the dots have replaced a lot of numbers: 14, 16, 18, etc. The set *D* is *infinite*, whereas every previous example was *finite*. The distinction between finite and infinite sets will get a lot of attention in Chapter 7.

We shall give one more example of an infinite set of numbers. In Chapter 2 we discuss *prime numbers,* and we offer a brief preview here. A positive integer is said to be *prime* if the only positive numbers that divide it are itself and 1. It is traditional for a number of reasons not to call 1 a prime. So the first prime is 2, the next is 3, then 5, 7, etc. The number 15 is not a prime because 15 is the product of 3 and 5.

Let *P* be the set of primes,

$$P = \{2, 3, 5, 7, 11, 13, 17, \ldots\}$$

A beautiful theorem to be proved in Chapter 2 asserts that this list goes on forever. In other words, there is an infinite number of primes; that is, the set *P* is infinite. We can easily give a definite procedure for carrying the list of primes just as far as we like, but there is no description (say, by a simple formula) that can help us to write the members of *P*.

Suppose that instead of just one set *A* we are looking at two sets: *A* and *B*. Then it is natural to try to compare the two sets, and we are led to the idea of *inclusion*. The set *A* is said to be included in *B* if *A* is part of *B* (allowing the possibility that *A* and *B* are the same). In other words, we say that *A* is included in *B* if each member of *A* is a member of *B*. In symbols, *x* ε *A* implies *x* ε *B*. Thus the set {2, 4} is included in each of the sets {2, 4, 7}, {2, 4, 11}, and {2, 4}, but it is not included in {2, 3, 7, 11}.

The notation ⊂ is in common use for inclusion, the phrase "*A* is included in *B*" thus being replaced by "*A* ⊂ *B*." We note again that equality of A and B is permitted when we write $A \subset B$. If in a given case it happens that $A \subset B$ and $A \neq B$ and we wish to write this compactly, we use the notation $A \subsetneq B$. In words, we say that A is *properly* included in B.

We ask: Given two sets A and B, are there any natural ways of forming new sets from them? There are in fact two ways that are especially important. They are called *intersection* and *union,* and the rest of this section will be devoted to discussing them.

The intersection of A and B, written $A \cap B$, is the set of all elements lying in *both* A and B. Here are three illustrations:

$$\{3, 5, 8\} \cap \{3, 6, 9\} = \{3\}$$
$$\{3, 5, 8\} \cap \{3, 7, 8\} = \{3, 8\}$$
$$\{3, 5\} \cap \{3, 5, 8\} = \{3, 5\}$$

Notice the special nature of the last example. If we write $A = \{3, 5\}$ and $B = \{3, 5, 8\}$, then we have $A \subset B$ and furthermore $A \cap B = A$. This is always the case: For any sets C and $D,$ if C is included in D, then the intersection of C and D is C. The converse is also true: If two sets C and D have the property $C \cap D = C,$ then $C \subset D$.

Consider again the example we gave earlier: the set of all men now living who were born in Philadelphia in 1941. Call this set A. By changing our point of view a little, we can write A as an intersection. Let B be the set of all men now living, and let C be the set of all people (men or women, dead or alive) who were born in Philadelphia in 1941. Then $A = B \cap C$.

Geometry offers interesting examples of intersections. Consider, for instance, a straight line $L,$ and let, in addition, a point be given. This point might be on L or it might not be on L. In Figure 1.1 point P is on L and point Q is not on L. A popular view in developing geometry is to say that L is nothing else than the set of all points P lying on L. (This point of view will be developed in detail in Chapter 5.)

Now suppose that two lines L and M are given. What can we say about their intersection? There are three possibilities. At one extreme, L and M might be the very same line, in which case $L \cap M = L$ (or equally well M). In the normal case (illustrated in Figure 1.2) L and M have exactly one point, say P, in common. Then $L \cap M = \{P\}$, the set whose only member is P. The third possibility occurs when lines L and M are parallel

$\bullet \, Q$

L —————————————————•————————————————
P

Figure 1.1

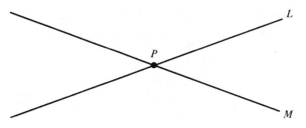

Figure 1.2

(Figure 1.3). Now we face a new puzzle in trying to determine what the intersection $L \cap M$ is. Or, to return to sets of numbers, what is $A \cap B$ if $A = \{2, 4, 7\}$ and $B = \{3, 5\}$? In either case the intersection must be a rather bizarre set, for plainly it can have no members at all.

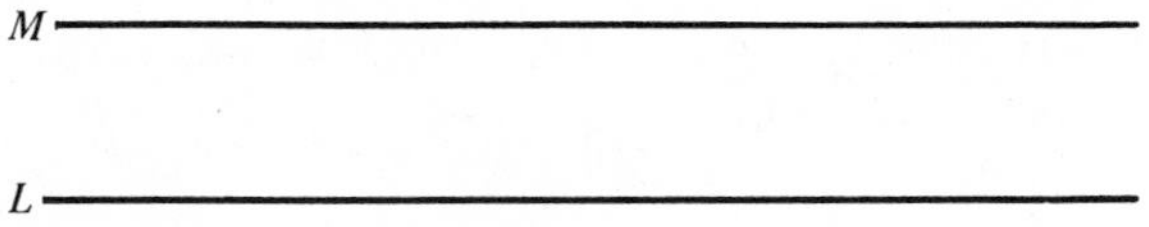

M ————————————————————

L ————————————————————

Figure 1.3

Now we would like to be able to say, without exception, that any two sets have an intersection. The way to solve this problem is to face it directly, and invent a set with no members. It is a good idea to have a name for this memberless set; words that have been used are *void* set, *empty* set, and *null* set. Let us agree on the term *null set*. As for notation, some authors have ventured to use 0, but there is some danger of confusion with the number zero. The notation ϕ has become fairly standard.

Note the following two properties of the null set: For any set A we have both $\phi \subset A$ and $\phi \cap A = \phi$.

As another geometric illustration consider a circle (the set of all points at a fixed distance from a given point called the center). Figure 1.4 shows the three possibilities for the intersection of a circle and a straight line: two points, one point, and the null set, respectively.

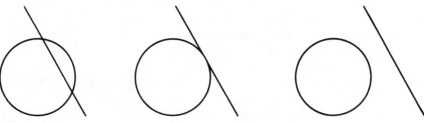

Figure 1.4

Given two sets A and B, there is another way of forming from them a third set which is, in a way, opposite to taking their intersection. We call it the union of A and B, write it $A \cup B$, and define it to be the set of elements that lie in A or in B or in both. In other words, $A \cup B$ is the set of elements belonging to at least one of A and B.

EXAMPLES

$\{2, 5, 7\} \cup \{2, 6, 11\} = \{2, 5, 6, 7, 11\}$

$\{2, 5, 7\} \cup \{3\} = \{2, 3, 5, 7\}$

$\{2, 5, 7\} \cup \{2\} = \{2, 5, 7\}$

$\emptyset \cup A = A$ for any set A.

To summarize, there are three concepts we have introduced in connection with two given sets: their union, their intersection, and inclusion of one in the other. They possess a surprisingly large number of properties. The simplest of these properties are the so-called commutative and associative laws for intersection and union. These are, respectively,

$A \cap B = B \cap A$

$(A \cap B) \cap C = A \cap (B \cap C)$

$A \cup B = B \cup A$

$(A \cup B) \cup C = A \cup (B \cup C)$

Formal proofs of these laws are easily constructed but are not very illuminating. In the next section and its exercises we shall see some deeper relations and interrelations concerning intersection and union.

Exercises

1. Describe the following sets verbally.
   (a) $S = \{$Washington, Adams, Jefferson, . . . , Johnson, Nixon$\}$
   (b) $S = \{$Ottawa, Rome, Paris, London, . . . , Washington, . . . $\}$
   (c) $S = \{$Mercury, Venus, . . . , Uranus, Neptune, Pluto$\}$
2. Describe the following sets in familiar terms.
   (a) $S = \{1, 4, 9, 16, 25, . . .\}$
   (b) $S = \{1, 3, 5, 7, 9, . . .\}$
   (c) $S = \{2, 3, 5, 7, 11, 13, 17, 19, 23, 29\}$
   (d) $S = \{17, 19, 23, 29, 31, 37, 41, 43, 47\}$
3. Exhibit the set of primes lying between 20 and 60. How many numbers are there in this set?
4. Let A be the set of all cities in the United States. Let B be the set of all cities in the United States with population 100,000 or more. Let C be the set of all cities in the United States with at least 500,000 women. Write all inclusions that hold among A, B, and C.

5. In addition to A, B, and C as in Exercise 4, let D be the set of those cities in the United States with at least 750,000 men. Find all possible inclusions among A, B, C, and D. (Hint: Do research on this if you like, but you may assume that the number of men and the number of women in an American city are roughly equal.)

6. A, B, and C are sets of positive integers. A has as members 1, 3, 5, 6; B has 3, 6; C has 3, 6, 7, 9, 10. Which of the following inclusions are true?

   (a) $B \subset A$

   (b) $A \subset B$

   (c) $A \subset C$

   (d) $B \subset C$

7. If A is the set of residents of the United States, B the set of people born before 1960, and C the set of Swiss citizens born since 1955, describe verbally $A \cap B \cap C$.

8. If A is the set of all American presidents prior to 1900, B is the set of all bachelors who ever lived, and C the set of all people who resided in North America since 1800, what is $A \cap B \cap C$? (For Americans only: Is $A \cap B \cap C$ empty? If not, who are its members?)

9. List the possibilities for the intersection of two circles. Draw a figure illustrating each.

10. Find:

    (a) $\{1, 2, 3, 4, \ldots\} \cap \{-1, -2, -3, 17, 11\}$

    (b) $\{0, 2, 4, 6, 8, \ldots\} \cap \{1, 3, 5, 7, \ldots\}$

11. Let $A = \{2, 4, 6, \ldots\}$ be the set of even positive integers, and let $B = \{3, 6, 9, \ldots\}$ be the set of those positive integers which are multiples of 3. Describe $A \cap B$.

12. The roster of major league players as of July 1, 1965 is examined, and the following three sets are formed: $A =$ the set of all major league pitchers, $B =$ the set of all National League pitchers, $C =$ the set of all American League pitchers. Give a relation between A, B, and C. What is $B \cap C$?

13. The sets $A = \{2, 5, 7, k\}$, $B = \{3, 5, 6, 8\}$, $C = \{5, 6\}$ are given, and you are told that $A \cap B = C$. What is k?

14. List the inclusion relations among $A \cap B$, $A \cap C$, $B \cap C$ where A is the set of all positive even numbers, B is the set of all positive multiples of 4, and C is the set of all prime numbers.

15. Describe $A \cup B$ where:

    (a) $A = \{0, 2, 4, \ldots\}$, $B = \{1, 3, 5, \ldots\}$

    (b) $A = \{-1, -2, -3, \ldots\}$, $B = \{0, 1, 2, 3, 4, \ldots\}$

    (c) $A = \{1, 4, 7, 10, \ldots\}$, $B = \{2, 5, 8, 11, \ldots\}$

16. If $A = \{1, 2, 3, 5, 9\}$, $B = \{4, 6, 7, 8\}$, and $C = \{1, 8, 9\}$, find:

   (a) $A \cap (B \cup C)$

   (b) $(A \cap B) \cup (A \cap C)$

   (c) Compare the results of (a) and (b).

17. If $A = \{1, 3, 4\}$, $B = \{3, 7, 8\}$, and $C = \{1, 3, 4, 6, 7\}$, find:

   (a) $(A \cup B) \cap C$

   (b) $A \cup (B \cap C)$

   (c) Compare the results of (a) and (b).

18. (a) Identify the set S, given the following information:

   $$S \cap \{3, 5, 8, 11\} = \{5, 8\}$$
   $$S \cup \{4, 5, 11, 13\} = \{4, 5, 7, 8, 11, 13\}$$
   $$\{8, 13\} \subset S$$
   $$S \subset \{5, 7, 8, 9, 11, 13\}$$

   (b) Show that no three of these conditions suffice to identify S.

2. Sets and Counting

In the discussion in the preceding section we introduced some formal operations for combining given sets to form new ones. We now want to establish a contact between these notions and something which is more familiar to all of us, counting.

First we need a symbol or abbreviation for the number of elements in the set S. Let $N(S)$ denote the number of elements in S. Thus, for instance, if S is the set of inhabitants of the city of Chicago, then $N(S)$ is the population of Chicago. If $S = \{1, -7, 13\}$, then $N(S) = 3$. A final example: If $S = \phi$, the null set, then $N(S) = 0$. At present this symbol $N(S)$ only makes sense for us if S is a finite set.

Starting with two finite sets S and T, we can form their union, $S \cup T$, and their intersection, $S \cap T$. From these we get four numbers $N(S)$, $N(T)$, $N(S \cup T)$, and $N(S \cap T)$, the number of elements in S, T, $S \cup T$, and and $S \cap T$, respectively. It is perfectly reasonable to expect that some relation exists among them. Our first objective is to determine this relation.

Without further ado we assert:

$$N(S \cup T) = N(S) + N(T) - N(S \cap T)$$

Having said this, all that remains is to prove it.

Given any element in $S \cup T$, it is of exactly one of the following three types:

1. in S but not in T
2. in T but not in S
3. in both S and T

We count up how many elements there are of each type. Since the elements of type 3 are precisely those in $S \cap T$, we have $N(S \cap T)$ elements of type 3. What about type 1? Any element is of type 1 if it is in S but not in

$S \cap T$. This number is clearly $N(S) - N(S \cap T)$. So, type 1 gives us $N(S) - N(S \cap T)$ elements. Similarly, type 2 gives us $N(T) - N(S \cap T)$ elements. Because among these three types we account for each element of $S \cup T$ exactly once,

$$N(S \cup T) = N(S \cap T) + (N(S) - N(S \cap T)) + (N(T) - N(S \cap T))$$
$$= N(S) + N(T) - N(S \cap T)$$

as claimed.

Let us see how this checks out in a particular case. Let $S = \{1, 3, 5, 7, 8\}$ and $T = \{1, 3, 6, 9\}$. Then $S \cup T = \{1, 3, 5, 6, 7, 8, 9\}$ and $S \cap T = \{1, 3\}$. Now, $N(S) = 5$, $N(T) = 4$, $N(S \cup T) = 7$ and $N(S \cap T) = 2$; substituting these values in the result that we have established, we have indeed that $7 = N(S \cup T) = 5 + 4 - 2 = N(S) + N(T) - N(S \cap T)$.

We put this result to use, now, to check on the consistency of data. Consider the following example.

In its advertising a tobacco company claims that, in testing 100 smokers (all of whom smoked cigarettes, a pipe, or both) they found that 70 smoked cigarettes, 60 smoked a pipe, and 20 smoked both. Do these figures make sense? Can they describe an actual situation?

We put this example into the framework of the discussion above. Let C be the set of cigarette smokers in the sample tested, and P the set of pipe smokers. From the data given we know:

1. $N(C \cup P) = 100$ (since each of the 100 people tested smoked cigarettes or a pipe)
2. $N(C) = 70$ (70 smoked cigarettes)
3. $N(P) = 60$ (60 smoked a pipe)
4. $N(C \cap P) = 20$ (20 smoked both)

By our result above, $100 = N(C \cup P) = N(C) + N(P) - N(C \cap P) = 70 + 60 - N(C \cap P)$. This gives $N(C \cap P) = 30$; hence, exactly 30 of those tested smoked both cigarettes and a pipe. This does not jibe with the figure given by the tobacco company, 20, hence the data they gave were not accurate.

Let us complicate the situation a little. Instead of talking about the union of two sets, we'll talk about the union of three (or more) sets. Previously we defined $A \cup B$, the union of two sets, as the set of elements that are in at least one of A or B. It is perfectly natural to define $A \cup B \cup C$ (the union of A, B, and C) as the set of all elements that are in at least one of A, B, or C. In fact, in this way we can define the union of any number of sets. Similarly, we defined $A \cap B$, the intersection of A and B, as the set of elements in both A and B. Clearly, by the intersection of A, B, and C, that is, $A \cap B \cap C$, we mean the set of elements that are in each of A, B, and C. This, too, can be extended to the intersection of any number of sets.

Suppose A, B, and C are three finite sets, with $N(A)$, $N(B)$, and $N(C)$ elements, respectively. What is $N(A \cup B \cup C)$? Before working out the answer precisely, let us see what one could expect to happen. If we simply count the elements in $A \cup B \cup C$ as $N(A) + N(B) + N(C)$, then we are over-counting elements which are in more than one of these sets. So we will have to adjust this simple-minded count by some compensating factors. For instance, if an element is in two of A, B, C but not in the third, then in $N(A) + N(B) + N(C)$ we will have counted it twice, so we must throw it out exactly once. This suggests that terms such as $-N(A \cap B)$, $-N(B \cap C)$, $-N(A \cap C)$ should be present in the answer. Let us try $N(A) + N(B) + N(C) - N(A \cap B) - N(B \cap C) - N(A \cap C)$ as our guess for $N(A \cup B \cup C)$. This still doesn't do the trick, for, consider an element that is in all of A, B, and C. In $N(A) + N(B) + N(C)$ it has been counted three times; in our compensating factors $-N(A \cap B)$, $-N(A \cap C)$, $-N(B \cap C)$ it has been thrown out three times. So, in the trial answer we wrote above, this element would not be counted at all. Thus, we should adjust the answer once more, putting such elements back in, with the term $N(A \cap B \cap C)$. Our guess at the result would then read:

$$N(A \cup B \cup C) = N(A) + N(B) + N(C) - N(A \cap B) - N(A \cap C)$$
$$- N(B \cap C) + N(A \cap B \cap C)$$

This is, in fact, a good guess at the answer. As a matter of fact, the discussion above which led to this is a perfectly sound proof. However, it will be instructive to give an alternate proof, which amounts to a formal reduction to the case of two sets.

Let $D = B \cup C$; then $A \cup B \cup C = A \cup D$. Hence

$$N(A \cup B \cup C) = N(A \cup D)$$
$$= N(A) + N(D) - N(A \cap D)$$

To finish the job we merely must know the value of the constituent pieces $N(D)$ and $N(A \cap D)$. The term $N(D)$ is easy, for $N(D) = N(B \cup C) = N(B) + N(C) - N(B \cap C)$ by our result for two sets. All that remains is $N(A \cap D)$. To handle $N(A \cap D)$ something a little more complicated, the *distributive law,* is needed. This law asserts that for any three sets X, Y, Z,

$$X \cap (Y \cup Z) = (X \cap Y) \cup (X \cap Z)$$

In order not to break the discussion, we leave its proof as an exercise (see Exercise 20). Since $D = B \cup C$, we have

$$A \cap D = A \cap (B \cup C) = (A \cap B) \cup (A \cap C)$$

by the distributive law. Thus

$$N(A \cap D) = N[(A \cap B) \cup (A \cap C)]$$
$$= N(A \cap B) + N(A \cap C) - N[(A \cap B) \cap (A \cap C)]$$

from the result for two sets. But of course

$$(A \cap B) \cap (A \cap C) = A \cap B \cap C$$

Hence

$$N(A \cap D) = N[(A \cap B) \cup (A \cap C)]$$
$$= N(A \cap B) + N(A \cap C) - N(A \cap B \cap C)$$

Plugging these values of $N(D)$ and $N(A \cap D)$ into the value of $N(A \cup B \cup C) = N(A \cup D)$, we get

$$N(A \cup B \cup C) = N(A) + [N(B) + N(C) - N(B \cap C)]$$
$$- [N(A \cap B) + N(A \cap C) - N(A \cap B \cap C)]$$
$$= N(A) + N(B) + N(C) - N(A \cap B) - N(A \cap C)$$
$$- N(B \cap C) + N(A \cap B \cap C)$$

Can you guess (or derive) the result for the union of four sets? For the union of any finite number of sets? Try to explain the alternation of the plus signs and minus signs in your answers.

Let us use the result for the union of three sets to examine some data. Suppose that in a poll made of 150 people the following information was obtained: 60 of them read the *New York Times,* 80 read *Harper's,* and 90 read *Playboy.* Furthermore, 50 read both the *New York Times* and *Harper's,* 20 read both the *New York Times* and *Playboy,* and 30 read both *Harper's* and *Playboy.* What, if anything, can we say about the number who read all three?

Let T be the set of those who read the *New York Times,* H the set of those who read *Harper's,* and P the set of those who read *Playboy.* Our data:

1. $N(T \cup H \cup P) \leq 150$ (the reason we say $\leq$ rather than $=$ is that it is possible that some person polled reads none of the three)
2. $N(T) = 60$
3. $N(H) = 80$
4. $N(P) = 90$
5. $N(T \cap H) = 50$
6. $N(T \cap P) = 20$
7. $N(H \cap P) = 30$

We want to know something about $N(T \cap H \cap P)$. Now,

$$150 \geq N(T \cup H \cup P) = N(T) + N(H) + N(P) - N(T \cap H)$$
$$- N(T \cap P) - N(H \cap P) + N(T \cap H \cap P)$$
$$= 60 + 80 + 90 - 50 - 20 - 30$$
$$+ N(T \cap H \cap P)$$

This gives $N(T \cap H \cap P) \leq 20$; that is, at most 20 of those polled read all three. Note that if we knew that each person polled reads at least one of the *New York Times, Harper's,* or *Playboy,* we could then give an exact answer, namely, that $N(T \cap H \cap P) = 20$.

We consider one more question of this type. It stems from one of the mathematical "riddles" posed by Lewis Carroll in *A Tangled Tale,* in the part entitled *Knot X.* In it, Clara and her aunt see a group of disabled military pensioners. Her aunt then poses the problem:

> Say that 70 per cent have lost an eye—75 per cent an ear—80 per cent an arm—85 per cent a leg—that'll do it beautifully. Now, my dear, what percentage, *at least* [italics Carroll's, not ours], must have lost all four?

> Lewis Carroll, the famous author of *Alice in Wonderland, Through the Looking Glass, The Hunting of the Snark,* and other wonderful works, was Charles Lutwidge Dodgson (1832–1898), a clergyman, Oxford don, and mathematician. His literary achievements far outstrip his mathematical ones, and although honored and remembered as a truly great writer, he is forgotten as a mathematician. He was very fond of mathematical puzzles and published a variety of them. In fact, *A Tangled Tale* is a series of such problems.

Let S be the set of pensioners, E the set of those that lost an eye, H those that lost an ear, A those that lost an arm, and L those that lost a leg. We suppose that $N(S) = 100$. We want an estimate for $N(E \cap H \cap A \cap L)$, the number of those unfortunates who lost all four members. Now, $S \supset E \cup H$, hence $100 = N(S) \geqq N(E \cup H) = N(E) + N(H) - N(E \cap H) = 70 + 75 - N(E \cap H)$. Hence $N(E \cap H) \geqq 45$. Similarly, $N(A \cap L) \geqq 65$. Therefore, since $S \supset (E \cap H) \cup (A \cap L)$,

$$100 = N(S) \geqq N[(E \cap H) \cup (A \cap L)]$$
$$= N(E \cap H) + N(A \cap L) - N(E \cap H \cap A \cap L)$$
$$\geqq 45 + 65 - N(E \cap H \cap A \cap L)$$

This gives $N(E \cap H \cap A \cap L) \geqq 10$. In short, at least 10 percent lost all four members. It is easy enough to construct an example satisfying the data given in which exactly 10 percent have lost all four members (see Exercise 22). Hence we obtained the best possible result.

Let us return to operations on sets. In counting the number of elements in $S \cup T$, we split the elements of $S \cup T$ into three types: those in S but not in T, those in T but not in S, and those in both S and T. This last type consists merely of the elements in $S \cap T$, a set operation that was introduced quite early. We now introduce a new set operation based on the description of the other two types of elements of $S \cup T$.

If S and T are any two sets, by the *difference* of S and T, which we write as $S - T$, we shall mean the set of elements that are in S but not in T. For instance, if S is the set of inhabitants of Chicago and T is the set of all females in the world, then $S - T$ is the set of all male inhabitants of Chicago. As with our two previous set operations—union and intersection—we can form new sets from old by means of the operation of differences. Moreover, with this new operation, as with the other two, certain algebraic relations hold. We shall pose several of these as exercises; a few of the more obvious ones we list now.

1. If $A \subset B$, then $S - B \subset S - A$.
2. $(S - T) \cap T = \phi$, the null set.
3. If $S - T = \phi$, then $S \subset T$.
4. $(S - T) \cup (S \cap T) = S$

Try to verify these.

Our interest, however, shall not be so much in the difference of any two sets but, rather, in a somewhat more special situation. Suppose that all the sets we shall discuss are subsets of a fixed set S. If A is such a subset of S, we call $S - A$ the *complement* of A in S. We write this as $C(A)$.

This construction (and symbol) enjoys some simple and nice properties. They are not too hard to prove and will be found in the exercises.

1. $C(C(A)) = A$. Note that this says that every subset of S is the complement in S of something; in fact, it is the complement of its own complement.
2. If $A \subset B$, then $C(B) \subset C(A)$. Notice that what this says is that forming complements reverses inclusion relations.

Having introduced the new operation—that of forming the complement—we might wonder how this new operation interacts with our two earlier operations, union and intersection. The complete story here is contained in the relations called *De Morgan's rules*. We first state the rules and then prove one of them, leaving the other as an exercise (see Exercise 16).

De Morgan's Rules: If A and B are subsets of S, then $C(A \cap B) = C(A) \cup C(B)$ and $C(A \cup B) = C(A) \cap C(B)$.

These rules express a certain symmetry, or, perhaps more accurately, an antisymmetry among C, $\cup$, and $\cap$. The application of C converts a union into an intersection and an intersection into a union.

We now prove the first of the De Morgan rules, namely, that $C(A \cap B) = C(A) \cup C(B)$. The usual strategy for showing the equality of two sets M and N is to show the two opposite inclusion relations $M \subset N$ and $N \subset M$. We shall use this approach here for $M = C(A \cap B)$ and $N = C(A) \cup C(B)$.

Since $A \cap B \subset A$, by one of the easy properties of C mentioned a moment ago, $C(A) \subset C(A \cap B)$. Similarly, $C(B) \subset C(A \cap B)$. Hence $C(A) \cup C(B) \subset C(A \cap B)$. Half the struggle is over—we have one of our desired inclusions—so on to the other half. We want to show that $C(A \cap B) \subset C(A) \cup C(B)$. Suppose that $x \ \varepsilon \ C(A \cap B)$; hence x is *not* in both A and B. Thus x fails to be in one of A or B. If x is not in A, it must be in $C(A)$; if x is not in B, it must be in $C(B)$. At any rate, x is in one of $C(A)$ or $C(B)$, hence x is certainly in $C(A) \cup C(B)$. Thus any $x \, \varepsilon \, C(A \cap B)$ is in $C(A) \cup C(B)$, which is to say, $C(A \cap B) \subset C(A) \cup C(B)$. Having established both $C(A) \cup C(B) \subset C(A \cap B)$ and $C(A \cap B) \subset C(A) \cup C(B)$, we have proved that $C(A \cap B) = C(A) \cup C(B)$, the desired result.

This interrelation among C, $\cup$, and $\cap$ goes over smoothly for any number of sets. Consider, for instance, $C(A \cap B \cap D)$. Looking at this as $C(E \cap D)$, where $E = A \cap B$, we get $C(E \cap D) = C(E) \cup C(D)$. But $C(E) = C(A \cap B) = C(A) \cup C(B)$. The net result of all this is that $C(A \cap B \cap D) = C(A) \cup C(B) \cup C(D)$. By the same argument we can show that $C(A \cap B \cap D \cap G) = C(A) \cup C(B) \cup C(D) \cup C(G)$. We can continue in this way to any finite number of sets; we have mentioned explicitly the case of four sets because we are about to make use of it in an alternate discussion of Lewis Carroll's puzzle about the disabled pensioners.

As before, let E be the set of those that lost an eye, H those that lost an ear, A those that lost an arm, and L those that lost a leg. The information given is that 70 percent of the pensioners are in E, 75 percent in H, 80 percent in A, and 85 percent in L. Hence $C(E)$ has 30 percent, $C(H)$ 25 percent, $C(A)$ 20 percent, and $C(L)$ 15 percent of the pensioners. Therefore $C(E) \cup C(H) \cup C(A) \cup C(L)$ has at most $30 + 25 + 20 + 15 = 90$ percent of the pensioners. But

$$C(E) \cup C(H) \cup C(A) \cup C(L) = C(E \cap H \cap A \cap L)$$

as we saw in the above discussion. Hence $C(E \cap H \cap A \cap L)$ has at most 90 percent of the pensioners; but then $E \cap H \cap A \cap L$ must have at least 10 percent of them. Thus at least 10 percent lost all four members.

It might be of some interest to read some of Carroll's comments on the various solutions that were sent to him by his readers. We give two quotes:

1. Solution (I adopt that of POLAR STAR, as being better than my own). Adding the wounds together, we get $70 + 75 + 80 + 85 = 310$, among 100 men; which gives 3 to each, and 4 to 10 men. Therefore the least percentage is 10.
2. BRADSHAW OF THE FUTURE and T.R. do the question in a piece-meal fashion—on the principle that the 70 per cent and the 75 per cent, though commenced at opposite ends of the 100, must overlap by *at least* 45 per cent; and so on. This is quite correct working, but not, I think, quite the best way of doing it.

Note that this second method he mentions, and deprecates, is the first solution we gave to the problem, and the POLAR STAR solution is essentially the second solution we gave.

In much of what has preceded, we have discussed subsets of a given set. One might easily wonder exactly how many different subsets a given finite set can have. We shall work this out very soon but, before going to the general situation, we take a close look at the first few cases.

If a set S_1 has one element, $S_1 = \{x_1\}$ say, then the only subsets of S_1 are the empty set, and S_1 itself. In other words, in this case S_1 has two distinct subsets. If a set S_2 has two elements, $S_2 = \{x_1, x_2\}$ say, then it is still easy to write out all the subsets of S_2. They are: ϕ, $\{x_1\}$, $\{x_2\}$, and $\{x_1, x_2\} = S_2$. Here we see that S_2 has four distinct subsets. Let us try one more case, that of a set having three elements. Let $S_3 = \{x_1, x_2, x_3\}$; the list of all the subsets of S_3 then is: ϕ, $\{x_1\}$, $\{x_2\}$, $\{x_3\}$, $\{x_1, x_2\}$, $\{x_1, x_3\}$, $\{x_2, x_3\}$, $\{x_1, x_2, x_3\} = S_3$. Here we note that S_3 has eight distinct subsets.

The answers we obtained can be expressed succinctly by the table below:

Number of elements	Number of subsets
1	2
2	4
3	8

Recognizing 2 as 2^1, 4 as 2^2, and 8 as 2^3, there seems to be a pattern forming in the first few cases. On the basis of this it might be reasonable to guess that a set having n elements has 2^n distinct subsets. Now, having an idea of what the answer might be, we set out to establish it.

Let S have n elements; $S = \{x_1, \ldots, x_n\}$, say. A subset A of S is completely described when we know which elements are in it and which are not. So we can describe A by a string of length n of IN's and OUT's, where in the ith place in this string we write IN if x_i, the ith element, is in A and OUT if it is not. (For instance, the subset $A = \{x_2, x_4\}$ of $S = \{x_1, x_2, x_3, x_4\}$ would be described by the string OUT, IN, OUT, IN.) The number of subsets of S thus coincides with the number of different strings or sequences of length n of IN's and OUT's. For each place in such a string we have two ways of filling it—either IN or OUT—and, since we can fill each place at will (that is, independently), the number of possible strings is

$$\underbrace{2 \cdot 2 \cdot 2 \cdots 2}_{n \text{ times}} = 2^n$$

There are, thus, 2^n such strings so, by the argument given above, there are 2^n different subsets of S. We shall treat this question again in Chapter 2, Section 4; there it will serve as an illustration of mathematical induction.

A word is in order about the last part of the argument given. What we are using is: If we can do something, Q, in q distinct ways and, independently, something else, R, in r distinct ways, then we can do Q and R together in qr distinct ways. A concrete way of visualizing this might be as follows. Suppose that Chicago and New York are connected by a series of one-way highways; q being available for the trip Chicago–New York and r for that of New York–Chicago. Then we can make the round trip between Chicago and New York in qr different ways. Why? Each route from Chicago to New York gives rise to r possible returns, hence r round trips. Because there are q routes from Chicago to New York, the number of different round trips is

$$\underbrace{r + r + \cdots + r}_{q \text{ times}} = qr$$

We shall treat counting problems with a similar flavor toward the end of the next section.

Exercises

1. What is $N(S)$ for the following sets S?
   (a) $S = \{\text{Mercury, Venus, } \ldots \text{, Uranus, Neptune, Pluto}\}$
   (b) $S = \{A, C, E, G, \ldots, W, Y\}$
   (c) $S = \{\text{Alabama, Alaska, Arizona, } \ldots \text{, Wisconsin, Wyoming}\}$
   (d) $S = \{\text{Alberta, British Columbia, } \ldots \text{, Quebec, Saskatchewan}\}$
2. What is $N(S)$ for the following sets S?
   (a) $S = L \cap M$ where L and M are parallel lines
   (b) $S = L \cap M$ where L and M are distinct nonparallel lines
   (c) $S = C \cap D$ where C and D are concentric circles of radius 1 and 2 respectively
   (d) $S = C \cap D$ where C is a circle, and D is a circle of the same radius having its center on the circumference of C
3. What is $N(S)$ for the following sets S?
   (a) $S = A \cup B$ where $A = \{-5, -4, 0, 1, 2\}$ and $B = \{-10, -9, \ldots, 0, 1, 2, \ldots, 10\}$
   (b) $S = C \cap D$ where $C = \{1, 3, 5, 7, \ldots\}$ and $D = \{-5, -3, 0, 1, 2, 3, 4, 5\}$
   (c) $S = E \cup F$ where E is the set of integers lying between 2 and 99 which are squares (of integers), and F is the set of integers lying between 2 and 99 which are cubes (of integers)
   (d) $S = E \cap F$ where E and F are as in (c)
   (e) $S = (G \cup H) \cap J$ where $G = \{-7, 0, -4, 2, 8\}$, $H = \{0, 2, 4, 6, 8\}$, and $J = \{-9, -4, 4, 6, 9\}$
4. Given that 70 percent of the population of North America have swum in the Atlantic Ocean, and that 60 percent have swum in the Pacific Ocean, what percentage at least have swum in both?
5. S and T are given sets.
   (a) If $N(T) = 8$, $N(S \cap T) = 5$, and $N(S \cup T) = 10$, what is $N(S)$?
   (b) If $N(S) = 9$, $N(T) = 6$, $N(S \cap T) = 4$, what is $N(S \cup T)$?

6. A survey revealed that 90 percent of Canadian homes have a radio and 80 percent have a TV set. What percentage, at least, have both a radio and a TV set?

7. S and T are finite sets. We are told that $N(S) = 7$, $N(S \cap T) = 4$. Which of the following are possible and which are impossible?
   (a) $N(T) = 7$, $N(S \cup T) = 7$
   (b) $N(T) = 10$, $N(S \cup T) = 13$
   (c) $N(T) = 13$, $N(S \cup T) = 10$
   (d) $N(T) = 3$, $N(S \cup T) = 6$

8. S and T are finite sets. We are told that $N(S \cap T) = 2$, $N(S \cup T) = 9$. Which of the following are possible and which are impossible?
   (a) $N(S) = 9$, $N(T) = 2$
   (b) $N(S) = 10$, $N(T) = 1$
   (c) $N(S) = 8$, $N(T) = 3$
   (d) $N(S) = 2$, $N(T) = 9$

9. Let a, b, c, d be integers ≥ 0 that satisfy $a \leq b$, $a \leq c$ and $b + c = a + d$. Prove that there exist sets S and T satisfying $N(S) = b$, $N(T) = c$, $N(S \cap T) = a$, $N(S \cup T) = d$.

10. Describe verbally the following sets:
    (a) $A - B$ where A is the set of all monarchs of England and B is the set of all married people who ever lived.
    (b) $(A - B) - C$ where A and B are as in (a) and C is the set of all males who ever lived.
    (c) $D - E$ where D is the set of all positive integers and E is the set of all even integers (positive, zero, or negative).

11. Precisely two of the following three sets A, B, C are equal. Which are they?

    $A = \{1, 2, 3, 4, 5, 6, 7\} - \{1, 3, 6, 12\}$
    $B = \{2, 3, 4, 5, 6, 7\} - \{1, 3, 9, 27\}$
    $C = $ the complement of $\{6, 9\}$ within the set $\{2, 4, 5, 6, 7, 9\}$

12. A half-open interval of real numbers is one which is "closed" on the left and "open" on the right. For example, the half-open interval from 1 to 2 consists of all real numbers x satisfying $1 \leq x < 2$, and the notation for it is $[1, 2)$. Prove that the intersection of $[3, 5)$ and $[4, 6)$ is again a half-open interval.

13. In this exercise complements are taken within the set of all integers. Verify the De Morgan rule $C(A \cup B) = C(A) \cap C(B)$ for the case $A = $ the set of all positive integers, $B = $ the set of all negative integers.

14. In this exercise, complements are taken within the set $\{1, 2, 3, \ldots, 19, 20\}$. Verify the De Morgan rule $C(A \cap B) = C(A)$

$\cup$ $C(B)$, where $A = \{1, 2, 3, \ldots, 14, 15\}$ and $B = \{9, 10, 11, \ldots, 19, 20\}$.

15. Let C denote complementation within a given set S. Prove, for any subsets A and B of S:
    (a) $C(C(A)) = A$
    (b) if $A \subset B$, then $C(B) \subset C(A)$

16. The first De Morgan rule, $C(A \cap B) = C(A) \cup C(B)$, was proved in the text. Prove the second De Morgan rule: $C(A \cup B) = C(A) \cap C(B)$.

17. Using the fact [(a) of Exercise 15)] that every subset of a given set S is a complement within S, show that each of De Morgan's rules implies the other.

18. Show that $S - (S - T) = S \cap T$.

19. Show that $C(A - B) = B \cup C(A)$.

20. Prove the distributive law $X \cap (Y \cup Z) = (X \cap Y) \cup (X \cap Z)$. (Hint: Try the strategy used on page 13 of the text in proving the first De Morgan rule. In other words, prove the two opposite inclusion relations for the sets $X \cap (Y \cup Z)$ and $(X \cap Y) \cup (X \cap Z)$. Note: There is a dual distributive law $X \cup (Y \cap Z) = (X \cup Y) \cap (X \cup Z)$. It can be proved in a similar way, or deduced from the first distributive law by De Morgan's rules as in Exercise 17.)

21. A public opinion poll shows that, when questioned about their reaction to three decisions of the government, 82 percent agreed with the government on the first decision, 74 percent on the second, and 77 percent on the third. What percentage, at least, agreed with the government on all three decisions?

22. Construct an example for Lewis Carroll's riddle in which exactly 10 percent of the pensioners lose an eye, ear, arm, and leg. (Hint: Number them off from 1 to 100; try to arrange it so that those from 1 to 10 lose all four.)

23. Given any integer n between 10 and 70, construct an example for Carroll's riddle in which n percent lose all four members.

24. Find an expression for $N(A \cup B \cup C \cup D)$ where A, B, C, D are finite sets.

25. How many subsets does S have if $S = A \cup B$ where $A = \{0, 1, -1, 2\}$ and $B = \{1, 2, 5\}$?

26. A is a set with five elements.
    (a) How many nonnull subsets does A have?
    (b) How many subsets does A have with exactly three elements?
    (c) Let B be a subset of A with two elements. How many subsets of A contain B? How many do not contain B?

27. (a) A is a set with $N(A) = k$. How many nonnull subsets does A have?
    (b) Suppose further that B is a subset of A with $N(B) = r$. How many subsets of A contain B?

28. Let A be any finite set and B a subset of A. Prove that the number of subsets of A containing B is the same as the number of subsets of A disjoint from B (i.e., intersecting B in the null set).

3. Functions

In what has preceded, we have discussed various aspects of set theory —the union and intersection of subsets of a given set, and the relation of these to counting. We'll now look at another side of set theory. Here we shall be concerned with mapping one set into another. Why? For one thing, as we shall see, it is a useful notion and leads to important results. Secondly, in one way or another, one could describe much of mathematics as a study of appropriate mappings or functions. In other words, the concept of mapping (or function) that we shall discuss will turn out to be both useful and central.

Let S and T be two sets; by a *function* or *mapping* from S to T we shall mean a rule or mechanism that assigns to each element of S an element of T. Before proceeding with some examples of functions and the development of some of the concepts about functions, we need some notation and vocabulary. Let f be a function from S to T; we shall often denote this by $f: S \to T$. If s is in S and t in T is that element to which s is assigned by the function f, then we shall denote this by writing $t = f(s)$. We often will call t the *image* of s under f. Sometimes we shall say that t is the value that f takes on at s, or that f *maps* s to t.

In the definition above, S and T could be any two sets; they need not have anything to do with mathematics but because our concern here is mathematics, for the most part we shall emphasize sets of numbers. However, in our discussion of the different types of functions, and in some of the examples, we make no such assumption about S and T. They can be any sets.

EXAMPLE 1. Let S be all the people living on Main Street in Midtown, Ohio and let T be the set of positive integers. If s is one of the residents of Main Street (i.e., $s \, \varepsilon \, S$), let $f(s)$ be the address of s on Main Street. So, if Joe Brown lives at 16 Main Street, $f(\text{Joe Brown}) = 16$. Note that since Joe's wife Mary (presumably) also lives at 16 Main Street, $f(\text{Mary Brown}) = 16$.

EXAMPLE 2. Again let S be the set of residents of Main Street in Midtown, Ohio and T the set of positive integers. Suppose that the Social Security system has been so perfected that each person—man, woman, and child—has a Social Security number and that no two people have the same Social Security number. Define the function $g: S \to T$ by $g(s) = $ the Social Security number of s.

Note that there is a difference in the nature of the two functions of Examples 1 and 2. As we saw in Example 1, f(Joe Brown) = f(Mary Brown) = 16; that is, two *different* elements in S can have the *same* image under f. Thus the values of f on the elements of S need not necessarily distinguish the elements of S. This is not the case in the second example. Here, two different elements in S have different Social Security numbers, so the values of the function $g: S \to T$ do distinguish between elements of S. This property of g, of taking on distinct values at distinct points of S, is a very important one. We call it "one-to-one"-ness; shortly we shall describe and define it more formally.

EXAMPLE 3.   Let S be the set of positive integers and let T be the set of positive odd integers. Let $h: S \to T$ be defined by $h(n) = 2n - 1$ for $n \; \varepsilon \; S$. Thus, for instance, $h(1) = 2(1) - 1 = 1$, $h(20) = 2(20) - 1 = 39$. This defines a function from S to T.

Note that this h, like the g in Example 2, does distinguish between elements of S. That is to say, if $n \neq m$, then $h(n) \neq h(m)$. Why? For, if $h(n) = h(m)$, then from $h(n) = 2n - 1$ and $h(m) = 2m - 1$ we would get that $2n - 1 = 2m - 1$; this would lead to the contradiction $n = m$.

However, the h of Example 3 enjoys a property that neither of the functions in Examples 1 or 2 does. We go into this now. Let x be any positive odd integer; then we can write x as $x = 2r - 1$ for some positive integer r. But then, $h(r) = 2r - 1 = x$. This says that *any* element of T appears as the image under h of some element of S. This property of h—let us call it "onto"-ness—is also a very important one, and we shall discuss it, too, in a short while. Note that the functions f and g in Examples 1 and 2 fail to satisfy this property. Clearly, we can find a positive integer large enough to exceed any address on Main Street or any Social Security number of the people on Main Street; such a large integer would then not be an image either in Example 1 or in Example 2.

EXAMPLE 4.   Let S be the set of real numbers and let T also be the set of real numbers. Let $f: S \to T$ be defined by $f(s) = s^2 + 6s + 7$. Thus, for instance, at $s = -4$, f takes on the value $f(-4) = (-4)^2 + 6(-4) + 7 = 16 - 24 + 7 = -1$. The function f that we are considering here has neither of the properties of "one-to-one"-ness nor "onto"-ness. To see the lack of the first of these properties, note that $f(-2) = (-2)^2 + 6(-2) + 7 = -1$, which we have seen is also the value of $f(-4)$. Therefore f does not distinguish, in the values it takes on, between -2 and -4. We claim that there is no $s \; \varepsilon \; S$ such that $f(s) = -10$; this is Exercise 11 at the end of the section. Hence -10 cannot arise as the image under f of any element of S. In consequence, f is not onto.

Example 4 is a very special case of a general family of functions from the set of reals to the set of reals, the family of functions known as *polynomials*. Let S and T both be the set of real numbers. Let $a_0, a_1, \ldots, a_n$ be any real numbers, with $a_0 \neq 0$. The function $f: S \to T$ defined on any $x \ \varepsilon \ S$ by $f(x) = a_0 x^n + a_1 x^{n-1} + \cdots + a_n$ is called a *polynomial* of degree n. The real numbers $a_0, a_1, \ldots, a_n$ are called the *coefficients* of the polynomial. The example we gave above, $f(x) = x^2 + 6x + 7$, is a polynomial of degree 2 with coefficients $a_0 = 1$, $a_1 = 6$, $a_2 = 7$. For instance,

$$f(x) = -\tfrac{11}{17}x^5 + 23x^3 + 7x^2 - \tfrac{9}{13}x + \tfrac{5}{7}$$

is a polynomial of degree 5; here the coefficients are given by $a_0 = -\tfrac{11}{17}$, $a_1 = 0$, $a_2 = 23$, $a_3 = 7$, $a_4 = -\tfrac{9}{13}$, and $a_5 = \tfrac{5}{7}$. Polynomials form an important class of functions and they will arise, from time to time, in what is to follow in the book.

EXAMPLE 5. Let S be any set and let $T = S$. In this case a function from S to T is really a function from S to itself. Define $i: S \to S$ by $i(s) = s$ for every $s \ \varepsilon \ S$. This function, which maps every element of S onto itself, plays a special role and so we give it a name: It is called the *identity function* on S. When we want to stress the role of S, we shall write the identity function on S as i_S.

Note, and we leave the easy details to the reader, that the identity function is endowed with both of the properties of "one-to-one"-ness and "onto"-ness.

EXAMPLE 6. Let S be the set of all integers $0, \pm 1, \pm 2, \ldots$, and T the set of positive integers. Define $f: S \to T$ by the rules:

1. $f(s) = 1$ if s is a negative integer
2. $f(0) = 101$
3. $f(s) = s$ if s is a positive integer

Thus, for instance, $f(-105) = 1$, $f(64) = 64$. Is the function f we defined one-to-one? Is it onto?

Now that we have seen some examples of functions and have pointed out some particular features that some functions may enjoy, it is best that we treat these aspects of functions a little more formally.

Definition. If S and T are sets, then the function $f: S \to T$ is said to be *one-to-one* if, given $s_1 \neq s_2$ in S, $f(s_1) \neq f(s_2)$.

Note that this says that the function f distinguishes between two distinct elements of S; that is, distinct elements in S end up, under f, in distinct images in T.

In the examples discussed above, only the functions in Examples 2, 3, and 5 are one-to-one.

We now define, more formally, the other property of functions that arose in our discussion of the examples.

Definition.   A function f from S to T is said to be *onto* if, given any $t_0 \; \varepsilon \; T$, we can find some element $s_0 \; \varepsilon \; S$ such that $t_0 = f(s_0)$.

So, a function f from S to T is onto T if every element in T crops up as the image under f of some element of S. In the examples discussed, only the functions in Examples 3, 5, and 6 are onto.

It may happen that for two given sets S and T there is a function f from S to T that is both one-to-one and onto. In that case f is called a *one-to-one correspondence* from S to T, and S and T are described as being in one-to-one correspondence (with each other). But wait! This last phrase is symmetric; it suggests a certain symmetry in the relation of being in one-to-one correspondence. However, the relation was defined in a rather unsymmetric way, namely, from the existence of a one-to-one correspondence *from S to T*. In this, S and T, at first glance, do not play symmetric roles. What is clearly needed is the following: If there is a one-to-one correspondence from S to T, then there is a one-to-one correspondence from T to S. We want to do this now; it leads to the very important idea of an *inverse* function.

So suppose that f is a one-to-one mapping of S onto T. Given any element $t \; \varepsilon \; T$, since f is onto, there is some element $s \; \varepsilon \; S$ such that $t = f(s)$. Moreover, since f is one-to-one, this element s is unique; that is, it is the one and only element in S with the property that $t = f(s)$. Thus there is no ambiguity whatsoever, once we are given t, in what element s will satisfy $t = f(s)$. Let this element s be called $g(t)$. This rule assigns to every element in T a unique element in $S,$ in other words, it defines a function g from T to S. This function is called the *inverse* of f and is usually written as f^{-1}. We emphasize that we have defined f^{-1} *only* when f is one-to-one and onto.

In rather vague words, what f^{-1} does is to "undo" f. This will be made more precise in a moment, when we will discuss the composition of functions. One might wonder whether or not f^{-1}, too, is one-to-one and onto (from T to S) when f is one-to-one and onto (from S to T). We could show this at once but it, also, will come out in the wash after the discussion of the composition of functions. Before doing this we consider a few examples of inverses of functions.

EXAMPLE 1.   Let S be the set of all integers. Define $f: S \to S$ by $f(s) = s - 6$. As is easily verified, f is one-to-one and onto. What is f^{-1}? Let $t \; \varepsilon \; S$; to know what $f^{-1}(t)$ is we have to find $x \; \varepsilon \; S$ such that

$f(x) = t$. But $f(x) = x - 6$; so in order to have $f(x) = t$ we would need to satisfy $t = f(x) = x - 6$. In other words, $x = t + 6$. Remembering that $x = f^{-1}(t)$ we have that $f^{-1}(t) = t + 6$. We therefore have an explicit formula for f^{-1}; $f^{-1}: S \to S$ is given by $f^{-1}(t) = t + 6$ for every $t \varepsilon S$.

EXAMPLE 2.   We consider the f of Example 3 discussed earlier. Recall that there S was the set of positive integers and T the set of positive odd integers. The function $f: S \to T$ defined by $f(s) = 2s - 1$ was seen to be one-to-one and onto. What is f^{-1}? To get $f^{-1}(t)$, where $t \varepsilon T$, we must find $s \varepsilon S$ such that $f(s) = t$. Using the explicit form of f, namely that $f(s) = 2s - 1$, we have $2s - 1 = t$, and so $s = (t + 1)/2$. Thus $f^{-1}(t) = (t + 1)/2$. Note that since t is odd, $t + 1$ is even, hence $(t + 1)/2$ is an integer, so is in S.

Note that in both examples above, in order to find $f^{-1}(t)$ we had to *solve*, in the relation defining f, for that value of s such that $f(s) = t$. This gave us $f^{-1}(t)$ in an explicit fashion. We shall see the same phenomenon arising more strikingly in the next example.

EXAMPLE 3.   Let R be the set of all real numbers. Consider the "alleged" mapping $f: R \to R$ defined by $f(x) = (x + 1)/(x - 2)$ for $x \varepsilon R$. For any $x \neq 2$, $f(x)$ can be computed from the formula to get a real number as an answer. For instance, $f(5) = (5 + 1)/(5 - 2) = 2$. However, something peculiar takes place at $x = 2$; if we tried to compute $f(2)$ from the formula we would get $f(2) = (2 + 1)/(2 - 2) = 3/0$, which is not defined (we cannot divide by 0). So f defines a perfectly decent function on R to R at all real numbers x other than $x = 2$. We avoid the trouble at $x = 2$ by not allowing it as a value of x. In other words, let $S =$ all real numbers x with $x \neq 2$. With this modification, f defines a function from S into R. Is it one-to-one? If not, for some $x \neq y$, $f(x) = f(y)$, which is to say, $(x + 1)/(x - 2) = (y + 1)/(y - 2)$. This yields $(x + 1)(y - 2) = (y + 1)(x - 2)$ which, when worked out, leads us to $3x = 3y$, and so $x = y$. In other words, f is indeed one-to-one. Is it onto? We claim no! For let $1 \varepsilon R$; we assert that $1 \neq f(s)$ for any $s \varepsilon S$. For if $1 = f(s) = (s + 1)/(s - 2)$, we would get $s - 2 = s + 1$ leading to the contradiction $3 = 0$. Hence 1 is *not* an image under f of any point in S. Let T be the set of all the real numbers x other than $x = 1$. Then f defines a function from S into T. We already know that f is one-to-one. It is a fact that f is onto T, as will become apparent in a moment.

Let us pretend that f indeed maps S onto T, so that f^{-1} exists. How would we go about finding an explicit formula for this hypothetical f^{-1}? If $t \varepsilon T$ and $s = f^{-1}(t)$, then, from the definition of $f^{-1}(t)$,

$f(s) = t$. But $f(s) = (s + 1)/(s - 2)$. Hence we have to solve for *that* s such that $(s + 1)/(s - 2) = t$. Therefore $s + 1 = (s - 2)t$, and so $s(t - 1) = 2t + 1$ giving us $s = (2t + 1)/(t - 1)$. In other words, $f^{-1}(t) = (2t + 1)/(t - 1)$. This formula gives us explicitly the s we need in order to see that f maps S onto T, and so the hypothetical f^{-1} has become the real thing. The formula also shows why $t = 1$ played a singular role and had to be excluded; if we try to put $t = 1$ in the formula $(2t + 1)/(t - 1)$ we find ourselves trying to divide by 0.

EXAMPLE 4.   Let $S = \{x_1, x_2, x_3, x_4\}$ and $T = \{y_1, y_2, y_3, y_4\}$ be two sets, each having four distinct elements. Define $f: S \to T$, explicitly, by declaring: $f(x_1) = y_2$, $f(x_2) = y_3$, $f(x_3) = y_4$, and $f(x_4) = y_1$. At sight we see that f is one-to-one and onto. What does $f^{-1}: T \to S$ look like? Since $f(x_4) = y_1$, we find $f^{-1}(y_1) = x_4$. Similarly, $f^{-1}(y_2)$ $= x_1$, $f^{-1}(y_3) = x_2$, and $f^{-1}(y_4) = x_3$.

We consider one more example.

EXAMPLE 5.   Let S be the set of all positive real numbers and let $T = S$. Define $f: S \to S$ by the rule $f(s) = 1/s$ for every $s \ \varepsilon \ S$. It is easy to verify that f is a one-to-one mapping of S onto itself. What is f^{-1}? We claim that, given $s \ \varepsilon \ S$, $f^{-1}(s) = 1/s$. Why? After all, if $f^{-1}(s) = t$, then t must be such that $f(t) = s$. But $f(t) = 1/t$; hence $1/t = f(t) = s$. This gives us $t = 1/s$. In short, $f^{-1}(s) = t = 1/s$, as claimed. Note the strange fact here that $f = f^{-1}$.

In the examples just discussed, we have seen sets in one-to-one correspondence with each other and the inverses of these one-to-one correspondences. We now return to the general concept of function; we want to introduce a method of combining two functions to get a third. This will give us an algebraic apparatus for operating rather fruitfully with functions.

Before doing the most general setup, we first discuss a special, but highly interesting, case. Let S be any set and let f and g be two mappings of S into itself. If $s \ \varepsilon \ S$, then $g(s)$ is an element of S; as an element of S, we can operate on $g(s)$ with the function f. This will give us an element $f(g(s))$ which is in S. We define the function $f(g(s))$ of S into S to be the *product* (or *composite*) of f and g. We shall often write the product of f and g as $f \circ g$ or merely fg.

The notion is a very important one, so we repeat what we have just said. If f, g are functions of S into S, then the function $f \circ g$ of S into S is defined by $(f \circ g)(s) = f(g(s))$ for every $s \ \varepsilon \ S$; that is, first apply g and, to the result of applying g, apply f.

Before continuing the discussion we consider two examples.

EXAMPLE 1. Let S be the set of real numbers and let $f: S \to S$ be defined by $f(s) = 5s + 6$ and $g: S \to S$ by $g(s) = 1/(s^2 + 1)$. What does $f \circ g: S \to S$ look like? From the recipe defining $f \circ g$,

$$(f \circ g)(s) = f(g(s)) = f\left(\frac{1}{s^2 + 1}\right) = 5\left(\frac{1}{s^2 + 1}\right) + 6 = \frac{5}{s^2 + 1} + 6$$

$$= \frac{6s^2 + 11}{s^2 + 1}$$

Note that $(f \circ g)(0) = (6 \cdot 0^2 + 11)/(0^2 + 1) = 11$. Of course we could also talk about $g \circ f$ in this setup. What does it look like? For $s \; \varepsilon \; S$,

$$(g \circ f)(s) = g(f(s)) = g(5s + 6) = \frac{1}{(5s + 6)^2 + 1}$$

Note that $(g \circ f)(0) = 1/((5 \cdot 0 + 6)^2 + 1) = 1/37$. Since $(f \circ g)(0) = 11$, whereas $(g \circ f)(0) = 1/37$, it is clear that $f \circ g \neq g \circ f$. Therefore it is possible that two functions from S into S need *not* "commute" with respect to the product of functions we have defined; that is, the order in which the product is taken makes a difference.

EXAMPLE 2. Let $S = \{x_1, x_2, x_3, x_4\}$. Let $f: S \to S$ be the function defined by $f(x_1) = x_2$, $f(x_2) = x_3$, $f(x_3) = x_4$, $f(x_4) = x_1$. Let $g: S \to S$ be defined by $g(x_1) = x_4$, $g(x_2) = x_2$, $g(x_3) = x_3$, $g(x_4) = x_1$. What does $f \circ g$ do to each element of S? We compute it:

$(f \circ g)(x_1) = f(g(x_1)) = f(x_4) = x_1$
$(f \circ g)(x_2) = f(g(x_2)) = f(x_2) = x_3$
$(f \circ g)(x_3) = f(g(x_3)) = f(x_3) = x_4$
$(f \circ g)(x_4) = f(g(x_4)) = f(x_1) = x_2$

Here, too, we can compute $g \circ f$; we get

$(g \circ f)(x_1) = g(f(x_1)) = g(x_2) = x_2$, $(g \circ f)(x_2) = x_3$, $(g \circ f)(x_3) = x_1$, $(g \circ f)(x_4) = x_4$

Note that in this case, too, $f \circ g \neq g \circ f$.

So far we have talked only about the composition of functions of S into S. But clearly we can speak about this in a wider context. We merely need the second function to map us *out* of the set *into* which the first one maps us. More precisely, we need to have three sets S, T, U, and functions $g: S \to T$ and $f: T \to U$. Then we can define $f \circ g: S \to U$ just as before, namely, by the rule $(f \circ g)(s) = f(g(s))$ for every $s \; \varepsilon \; S$. Note that in this setup it does not even make sense to speak about $g \circ f$ unless $S = U$. Why? Since f carries T into U, if $g \circ f$ made sense, g would have to act on U. But g is defined on S by hypothesis. Thus we would need $S = U$. Note, furthermore, that even to ask whether $f \circ g = g \circ f$ would require that $S = T = U$.

Let $f: S \to T$ be a one-to-to-one mapping of S onto T. Hence, as we have seen, we can define the inverse of f, which we called f^{-1}; it is a mapping of T into S. It might be of interest to try out the product we have just defined on these two functions f and f^{-1}. The result will explain the earlier remark that f^{-1}, in some sense, "undoes" f. We compute $f^{-1} \circ f$. If $s \, \varepsilon \, S$, then $(f^{-1} \circ f)(s) = f^{-1}(f(s))$. However, by the very definition of f^{-1}, if we let $t = f(s)$, then $f^{-1}(t) = s$. In other words, $f^{-1}(f(s)) = s$. But then $(f^{-1} \circ f)(s) = s$ for every $s \, \varepsilon \, S$. This says that $f^{-1} \circ f = i_S$, the identity map of S onto itself. We leave it to the reader to verify that $f \circ f^{-1} = i_T$, the identity map of T onto itself.

These two relations, $f^{-1} \circ f = i_S$ and $f \circ f^{-1} = i_T$, make it easy to see that f^{-1} is a one-to-one mapping of T onto S, as promised earlier. Why? First, the "one-to-one"-ness. Suppose that $f^{-1}(t_1) = f^{-1}(t_2)$ with $t_1, t_2 \, \varepsilon \, T$; then $f(f^{-1}(t_1)) = f(f^{-1}(t_2))$. But this translates into $(f \circ f^{-1})(t_1) = (f \circ f^{-1})(t_2)$; that is, $i_T(t_1) = i_T(t_2)$ (since $f \circ f^{-1} = i_T$). However, $i_T(t_1) = t_1$ and $i_T(t_2) = t_2$; the net result of all this is that $t_1 = t_2$. Therefore f^{-1} is indeed one-to-one. Why is it onto? Let $s \, \varepsilon \, S$; we want to exhibit some element $t \, \varepsilon \, T$ such that $s = f^{-1}(t)$. We do it explicitly. Let $t = f(s)$; then $f^{-1}(t) = f^{-1}(f(s)) = (f^{-1} \circ f)(s) = i_S(s) = s$. Hence f^{-1} is onto S.

These identity maps i_S and i_T have some further significant algebraic properties. Let f be any map of S into T and let i_T be the identity map on T. Then we have $f: S \to T$ and $i_T: T \to T$ and so we can speak about $i_T \circ f$. What is it exactly? If $s \, \varepsilon \, S$, then $(i_T \circ f)(s) = i_T(f(s))$. However, $i_T(t) = t$ for every $t \, \varepsilon \, T$, and, since $f(s) \, \varepsilon \, T$, $i_T(f(s)) = f(s)$. In short, $(i_T \circ f)(s) = f(s)$ for every $s \, \varepsilon \, S$. But this says that $i_T \circ f = f$. Similarly, we can show that $f \circ i_S = f$, where i_S is the identity map on S (see Exercise 16). In particular, when $S = T$, $i_S = i_T$; the relations above then tell us that for any map f of S into itself, $i \circ f = f \circ i = f$, where $i = i_S$.

Suppose we have the situation $g: S \to T$, $f: T \to U$. Hence we can talk about $f \circ g: S \to U$. Two questions present themselves naturally.

1. If f and g are both one-to-one, is $f \circ g$ then also one-to-one?
2. If f and g are both onto, is $f \circ g$ then also onto?

We show that the answer to each one of these questions is yes.

Suppose, then, that both of the functions $g: S \to T$ and $f: T \to U$ are one-to-one. Suppose that for some $s_1, s_2 \, \varepsilon \, S$, $(f \circ g)(s_1) = (f \circ g)(s_2)$. We want to show that this forces $s_1 = s_2$. Now $f(g(s_1)) = (f \circ g)(s_1) = (f \circ g)(s_2) = f(g(s_2))$. Since f is one-to-one, and since $f(g(s_1)) = f(g(s_2))$, we conclude that $g(s_1) = g(s_2)$. However, since g is one-to-one, this last relation gives us $s_1 = s_2$, as desired. Hence $f \circ g$ is one-to-one. This answers question 1.

Now to question 2. Suppose that both functions $g: S \to T$ and $f: T \to U$ are onto. We want to show that every $u \, \varepsilon \, U$ is an image of some element of S under $f \circ g$. Given $u \, \varepsilon \, U$, we have $u = f(t_0)$ for some appropriate $t_0 \; \varepsilon \; T$, since $f: T \to U$ is onto. Now, since $g: S \to T$ is onto, t_0 is the image under g of some element $s_0 \; \varepsilon \; S$; that is, $t_0 = g(s_0)$. But then $(f \circ g)(s_0) = f(g(s_0)) = f(t_0) = u$. We have produced an element $s_0 \; \varepsilon \; S$ such that $(f \circ g)(s_0) = u$. Therefore $f \circ g$ is onto. This answers question 2.

We combine these last two results to obtain: If $g: S \to T$ and $f: T \to U$ are both one-to-one and onto, then $f \circ g: S \to U$ is also one-to-one and onto. This observation will be of great importance to us in Chapters 3 and 4, when we discuss groups. There we shall be working with just one set S, and the observation will take the following form: If f, g are one-to-one correspondences of S onto itself, then $f \circ g$ is a one-to-one correspondence of S onto itself.

We close this section with some remarks interrelating mappings and counting. We ask two questions:

1. Let S be a finite set having n elements. How many distinct mappings are there of S into itself?
2. Let S be a finite set having n elements. How many distinct one-to-one mappings are there of S onto itself?

Before considering the general case, let us look at the situation where S is small. Perhaps this will give a hint of what the story will be in general.

When S has only one element, say x_1, then everything is easy. If f is any mapping of S into S, it must send x_1 somewhere. Where? Since S has only the one element x_1, this element has nowhere to go under f but into x_1. In short, $f(x_1) = x_1$. Thus S has only one mapping, the identity map; this mapping also happens to be one-to-one and onto.

Suppose, next, that S has two elements, x_1 and x_2. A mapping of S into itself can send x_1 either into x_1 or x_2. The mappings which send x_1 into x_1 are:

$$i: x_1 \to x_1 \quad \text{and} \quad f_2: x_1 \to x_1$$
$$x_2 \to x_2 \qquad\qquad\quad x_2 \to x_1$$

On the other hand, the mappings that send x_1 into x_2 are:

$$f_3: x_1 \to x_2 \quad \text{and} \quad f_4: x_1 \to x_2$$
$$x_2 \to x_2 \qquad\qquad\quad x_2 \to x_1$$

Thus S has four distinct mappings into itself. Glancing at the list, we see that only two of these mappings—i and f_4—are one-to-one and onto. Thus S has two one-to-one mappings onto itself.

Now to the story in general. Let $S = \{x_1, \ldots, x_n\}$. If $f: S \to S$ is a function of S into itself, to know f we need to know only where it sends each of $x_1, \ldots, x_n$. Now we can map x_1 into any of n possible images (namely, any of $x_1, \ldots, x_n$). Similarly, we can send x_2—independently of where we sent x_1—into any one of n possible images. Likewise for each x_i. As we saw at the end of Section 2, since we have n possible images for each of $x_1, \ldots, x_n$, we have

$$\underbrace{n \, n \cdots n}_{n \text{ times}} = n^n$$

possible ways of mapping S into itself. Hence there are n^n mappings of S into itself.

The story for the one-to-one mappings is considerably different. We no longer have the same freedom as we had above. Let us argue the case $n = 3$ first. Here we already have all the features of the general situation, but we have the advantage that we can see everything that is going on in a very explicit manner.

Let $S = \{x_1, x_2, x_3\}$. If f is a one-to-one mapping of S onto itself, f sends x_1 to some element of S. We have three possible choices for the image of x_1 under f. However, once an image has been chosen for x_1, it *cannot* be the image of x_2 (or any other x_i) under f because f is one-to-one. So we would have only two choices left for the image of x_2 under f. Once the choice of the images of x_1 and x_2 has been made, two elements of S have been used up and are no longer available as the image of x_3 under f (again since f is one-to-one). Hence for x_3 we have only one image possible, once we have chosen the images for x_1 and x_2. Thus, in an argument similar to that given at the end of Section 2, we have $3 \cdot 2 \cdot 1 = 6$ possible choices for f.

Let us see what we just said somewhat more schematically.

Image of x_1	Possible f's			
x_1	$x_2 \to x_2$	$x_2 \to x_3$		
	$x_3 \to x_3$	$x_3 \to x_2$		
x_2	$x_2 \to x_1$	$x_2 \to x_3$		
	$x_3 \to x_3$	$x_3 \to x_1$		
x_3	$x_2 \to x_1$	$x_2 \to x_2$		
	$x_3 \to x_2$	$x_3 \to x_1$		

Notice that for each possible image of x_1 we find two distinct one-to-one mappings of S onto itself.

What happens in the general case? If $S = \{x_1, \ldots, x_n\}$ and f is a one-to-one mapping of S onto itself, f can send x_1 into any one of n choices; however, once it sends x_1 somewhere, by the "one-to-one"-ness, we are left with only $n - 1$ possible places to map x_2. For x_3 we then get $n - 2$ possible images, and so on. Hence the number of possible f's (i.e., one-to-one mappings of S onto itself) is given by $n(n - 1)(n - 2) \ldots 2 \cdot 1$. This number, which we write as $n!$ and which we call n factorial, will appear often in Chapters 2, 3, and 4. Note some early values of $n!$: $1! = 1$, $2! = 2$, $3! = 6$, $4! = 4 \cdot 3 \cdot 2 \cdot 1 = 24$, $5! = 5 \cdot 4 \cdot 3 \cdot 2 \cdot 1 = 5 \cdot 4! = 120$.

The set of all the one-to-one correspondences of a set S, having n

elements, is an important mathematical object and has been studied a great deal. As we have seen, it has $n!$ elements. We shall take a much closer look at it in Chapter 3.

One final remark. We have spoken about the one-to-one mappings of a finite set *onto* itself. However, for finite sets there is a redundancy in talking about one-to-one onto. It is automatic that any one-to-one mapping of a finite set *into* itself is already *onto*. Similarly any onto mapping of a finite set is automatically one-to-one. This remark reappears in Chapter 2 under the guise of the Pigeon-Hole Principle.

Exercises

1. Let $S = T$ be the set of human beings alive today. Determine which of the proposed $f: S \to T$ define functions from S to T.
   (a) $f(s) =$ father of s
   (b) $f(s) =$ oldest living sister of s
   (c) $f(s) =$ oldest living member of the immediate family of s. (The immediate family consists of: the person, brothers and sisters, and parents.)
2. Let $S = T$ be the set of all human beings that ever lived. Determine which of (a), (b), (c) define functions from S to T.
3. In Exercise 2(a), how would you describe $(f \circ f)(s)$?
4. Let S be the set of all presidents of the United States and let T be the set of integers. Let f assign to each president the year of his first inauguration. Find $f(x)$ where
   (a) $x =$ John F. Kennedy
   (b) $x =$ Richard M. Nixon
   (c) $x =$ Abraham Lincoln
5. Let S be the set of positive integers and $T = \{0, \Box\}$. Define $f: S \to T$ by $f(x) = \Box$ if x is a perfect square and $f(x) = 0$ otherwise. Compute $f(1)$, $f(100)$, $f(105)$, $f(99)$, $f(5041)$.
6. Let $S = T$ be the set of all real numbers. Which of the following define functions from S to T?
   (a) $f(s) = \dfrac{1}{11} s + 13$
   (b) $f(s) = \dfrac{1}{11} s + 13 + \dfrac{16}{s+1}$
   (c) $f(s) = \sqrt{s}$
   (d) $f(s) = \sqrt[3]{s}$
   (e) $f(s) = \dfrac{6s + 5}{s + 5}$ if $s \neq -5$
        $f(-5) = \dfrac{6}{7}$
   (f) $f(s) = -6$
   (g) $f(s) = s^2 + 1$ if $s < 0$

$$f(s) = s + 1 \quad \text{if } s \geqq 0$$
(h) $f(s) = s \quad \text{if } s \geqq 0$
$$f(s) = -s \quad \text{if } s < 0$$
(i) $f(s) = 0 \quad \text{if } s < -10$
$$f(s) = 14 \quad \text{if } s = -10$$
$$f(s) = \frac{1}{s + 10} \quad \text{if } s > -10$$

7. Find the f's in Exercise 6 which do define functions from S to T, determine which are one-to-one, which are onto, and which are both.

8. Let $S = T$ be the set of all positive real numbers. Which of the following define functions from S to T?

(a) $f(s) = \dfrac{1}{11} s + 13$

(b) $f(s) = \dfrac{1}{11} s + 13 + \dfrac{16}{s + 1}$

(c) $f(s) = \sqrt{s}$

(d) $f(s) = \dfrac{6s + 5}{s + 5}$

(e) $f(s) = -6$

9. In Exercise 8, for those which do define functions from S to T, determine if they are one-to-one, onto, or both.

10. Show that the function defined by $f(x) = x^2 + 5x + 6$ from the set of real numbers to the set of real numbers is neither one-to-one nor onto.

11. Show that the function $f: S \to S$ defined from S, the set of real numbers, to itself by $f(x) = x^2 + 6x + 7$ never takes on the value -10.

12. If a and b are real numbers, show that the function defined by $f(x) = x^2 + ax + b$ for every real number x, is neither one-to-one nor onto as a function from the set of real numbers to the set of real numbers.

13. Show that the function, defined by $f(x) = x^3$, from the set of real numbers to the set of real numbers, is one-to-one and onto. What is f^{-1}?

14. (a) Show that if S is the set of positive real numbers, then $f: S \to S$, defined by $f(x) = (x + 1)/(x + 2)$ for $x \varepsilon S$, is a one-to-one mapping from S to S.

    (b) Is f onto?

15. Let S be the set of rational numbers. Suppose that $f: S \to S$ is defined by:

$f(1) \quad = -1$
$f(-1) = 1$
$f(s) \quad = s \quad \text{if} \quad s \neq 1, s \neq -1$

Show that f is one-to-one and onto, and find f^{-1}.

16. If S and T are any sets and $f: S \to T$, show that $f \circ i_S = f$.
17. (a) Let u be a fixed element of a set S. Define $h: S \to S$ by $h(s) = u$ for every $s \in S$. Prove that $h \circ f = h$ for every function $f: S \to S$.
    (b) Let $k: S \to S$ be such that $k \circ f = k$ for all $f: S \to S$. Prove that there exists an element $v \in S$ such that $k(s) = v$ for every $s \in S$.
18. Let S be the set of positive real numbers and let $f: S \to S$ be defined by $f(s) = 1/s$ for $s \in S$. Prove that $f \circ f = i_S$.
19. If $f: S \to S$ is a mapping such that $f \circ f = i_S$, prove that f is one-to-one and onto. What is f^{-1}?
20. Let S be the set of real numbers. For the functions $f, g: S \to S$ defined below, find $f \circ g$ and $g \circ f$.
    (a) $f(s) = s$ if $s \geq 0$
        $f(s) = -s$ if $s < 0$
        $g(s) = s^2$
    (b) $f(s) = \dfrac{s^2 + 2}{s^2 + 1}$
        $g(s) = \dfrac{s^2 + 1}{s^2 + 2}$
    (c) $f(s) = 7s - 6$, $g(s) = \dfrac{1}{7} s + \dfrac{6}{7}$
    (d) $f(s) = 6$, $g(s) = -7$
21. Let $S, T, U,$ and V be sets and $h: S \to T$, $g: T \to U$, and $f: U \to V$. Prove the *associative law* $f \circ (g \circ h) = (f \circ g) \circ h$.
22. Let $f: S \to T$ be a one-to-one correspondence. Suppose that $g: T \to S$ is such that $g \circ f = i_S$ and $f \circ g = i_T$. Prove that $g = f^{-1}$. (This proves the *uniqueness* of the inverse of a one-to-one correspondence.)
23. If $f: S \to T$ is a one-to-one correspondence, prove that $(f^{-1})^{-1} = f$.
24. If $g: S \to T$ and $f: T \to U$ are one-to-one correspondences, prove that $(f \circ g)^{-1} = g^{-1} \circ f^{-1}$.
25. We say that $h: T \to S$ is a *right inverse* of $f: S \to T$ if $f \circ h = i_T$. Show that if f has a right inverse, then f must be onto.
26. Show that if $g: S \to T$ and $f: T \to U$ both have right inverses, then $f \circ g$ has a right inverse.
27. If $f: S \to T$ has a right inverse and if h, k are mappings from $W \to S$ such that $h \circ f = k \circ f$, then $h = k$.
28. We say that $f: S \to T$ has a *left inverse* $h: T \to S$ if $h \circ f = i_S$. Show that if f has a left inverse, then f is one-to-one.
29. Show that if $g: S \to T$ and $f: T \to U$ both have left inverses, then $f \circ g$ has a left inverse.
30. If $f: S \to T$ has a left inverse and if v, w are mappings from T to U such that $f \circ v = f \circ w$, show that $v = w$.

31. Let S be the set of all real numbers and T the set of positive reals. Show that the mapping $f: S \rightarrow T$ defined by $f(s) = s^2$ has a right inverse. Is this right inverse unique?

32. Let S be the set of positive reals and T the set of all reals. Define $f: S \rightarrow T$ by $f(s) = \sqrt{s}$. Show that f has a left inverse. Is it unique?

33. If $f: S \rightarrow T$ has both a left inverse g and a right inverse h, show that f is one-to-one and onto. Moreover, show that $g = h = f^{-1}$.

34. Let S be a nonempty set. Show that if $f_0: S \rightarrow S$ is such that $f_0 \circ f = f \circ f_0$ for all mappings f of S into itself, then $f_0 = i_S$.

35. Suppose that S has more than two elements. Show that if $f_0: S \rightarrow S$ is a one-to-one mapping of S onto S such that $f_0 \circ f = f \circ f_0$ for all one-to-one mappings of S onto S, then $f_0 = i_S$.

36. Let S be the set of all positive integers. Let $f: S \rightarrow S$ be defined by $f(1) = 2$, $f(2) = 3$, $f(3) = 1$, $f(s) = s$ if $s > 3$. Show that $(f \circ f) \circ f = i_S$. What is f^{-1}?

37. Let $S = \{x_1, x_2, x_3, x_4\}$. Find $f \circ g$ and $g \circ f$ where f, g are the mappings of S into S given by:
    (a) $f(x_1) = x_2$, $f(x_2) = x_3$, $f(x_3) = x_4$, $f(x_4) = x_1$, and $g(x_1) = x_1$, $g(x_2) = x_2$, $g(x_3) = x_4$, $g(x_4) = x_3$
    (b) $f(x_1) = x_4$, $f(x_2) = x_2$, $f(x_3) = x_3$, $f(x_4) = x_1$, and $g = f$
    (c) $f(x_1) = f(x_2) = f(x_3) = f(x_4) = x_1$, and $g(x_1) = x_3$, $g(x_2) = x_1$, $g(x_3) = x_3$, $g(x_4) = x_4$

38. Let S and T be finite sets having m and n elements respectively. How many mappings are there from S into T?

39. Let S and T be finite sets having m and n elements respectively. If $m \leq n$, how many one-to-one mappings are there of S into T?

Chapter 2
Number Theory

So far the discussion has been about a rather abstract realm of things. We have considered sets whose objects are, by and large, nameless and faceless; just objects. Yet, even in the presence of such anonymity, we saw that much could be said. We now propose to make a rapid descent from the highly abstract to the very concrete. Instead of discussing any set, we fix our attention on one set well known to us all, the set of integers.

To make sure that we are talking about the same thing, we shall mean by the set of integers the collection of all numbers $0, \pm 1, \pm 2, \ldots$.

In the first part of this chapter we shall deal rather informally with many properties of the integers. Many of the things that we use we shall actually prove later on in the chapter, but at the beginning we operate with these as well-known facts. We can get formal later.

All right, what aspects of the integers shall we consider first? Initially we'll talk about factorization of integers and the fundamental pieces that enter into such a factorization—the prime numbers. These numbers have fascinated mathematicians, and most human beings, since time immemorial. Part of their fascination is that it is so easy to come up with questions about them and so hard to answer these questions. All considered, however, whatever is known about the set of primes (and we shall soon see that much is known and a great deal more unknown) attests to the power of a large variety of mathematical thinking and technique.

1. Prime Numbers

In this section we shall give very few proofs; instead, we shall discuss prime numbers, their properties, results known about them, and results as yet unknown. We are all familiar with the notion of such numbers, namely, those that cannot be factored in a nontrivial manner. More precisely:

Definition.   A positive integer $p \neq 1$ is said to be a *prime number* (or a *prime*) if it is impossible to write $p = ab$ where a, b are positive integers such that $1 < a < p, 1 < b < p$.

The exclusion of 1 from the list of primes is for technical reasons.

Experimenting with small numbers we are able to write some primes: 2, 3, 5, 7, 11, 13, ... Some immediate questions come to mind. Does this dot, dot, dot go on forever? More formally, is there an infinity of primes? How would we go about finding all the primes less than 100,000, say, in a practical way? How are the primes distributed amongst the integers? Is there a "formula" that gives us the prime numbers? How are the integers built up from the primes? and a host of others.

We begin with a simple observation. Given a positive integer $a > 1$, then a is divisible by some prime. What do we mean by divisible? We define, for any two integers a, b (positive, negative, or 0), that a is *divisible* by b if there exists an integer c such that $a = bc$. For convenience we write this as $b|a$; we denote that a is not divisible by b (or b does not divide a, or b is not a factor of a) by $b\nmid a$. Negative numbers are fully admitted in determining divisibility. For instance, $(-3)|6$ since $6 = (-3)(-2)$.

The following are not difficult to verify:

1. For any integer a, $1|a$.
2. For any integer b, $b|0$.
3. If $a|b$ and $b|c$, then $a|c$.
4. If $a|b$ and $a|c$, then $a|(xb + yc)$ for any integers x and y.
5. If $a|1$, then $a = 1$ or $a = -1$.
6. If $a|b$ and $b|a$, then $a = b$ or $a = -b$.
7. If a and b are positive and $a|b$, then $b \geqq a$.
8. For any integer a, $a|a$.

As a sample of how one proceeds to prove the assertions above let us consider item 4, which claims that if $a|b$ and $a|c$, then $a|(xb + yc)$ for all integers x and y. Since $a|b$, $b = au$ for some integer u; similarly, since

$a|c$, $c = av$ for some integer v. Now

$$xb + yc = x(au) + y(av) = a(xu + yv)$$

Hence, by the definition of divisibility, we get $a|(xb + yc)$. Two more of these assertions appear in Exercise 3.

We return to the statement that initiated this discussion of divisibility: Given an integer $a > 1$, then a is divisible by some prime. Let A be the set of integers b with $b > 1$ and $b|a$; A has a finite number of elements for if $b|a$, then $a \geq b$. By familiar properties of the integers, or more formally, by assumption L in Section 4, A has a smallest element b_o. We claim that b_o is a prime. For suppose that c_o divides b_o, where $1 < c_o < b_o$. Then, since $c_o|b_o$ and $b_o|a$, we have $c_o|a$. Thus $c_o \varepsilon A$, which is impossible since $c_o < b_o$ and b_o was the smallest element in A. Therefore b_o is indeed a prime; since $b_o|a$ our assertion has been verified.

Euclid first showed that there exist infinitely many primes. His proof is simple. Suppose there is only a finite number of primes; we enumerate these as $p_1, p_2, \ldots, p_n$. Let $a = 1 + p_1 p_2 \cdots p_n$. Now $a > 1$ and so a is divisible by some prime. Clearly a is not divisible by p_1, for otherwise $p_1|1$ would follow, resulting in $p_1 = 1$; but 1 was ruled out of the family of primes. Similarly, $p_i \nmid a$ for $i = 2, 3, \ldots, n$. This contradiction—a is divisible by some prime and yet is not divisible by any of $p_1, \ldots, p_n$ which enumerate all the primes—proves the existence of an infinite number of primes.

Let us refine this process a little. Given any integer a, on division by 4 it leaves a remainder of 0, 1, 2, or 3. Thus we can write a as $a = 4n + b$ where $b = 0, 1, 2,$ or 3, and n is some integer. This is true even if a is negative; for example, $-7 = 4(-2) + 1$. In order that a be a prime, clearly $b \neq 0$, otherwise a would be divisible by 4. If $b = 2$, then $a = 4n + 2 = 2(2n + 1)$; if a, in addition, is a prime, then it is positive, and since we have factored a as $a = 2(2n + 1)$ we must have $2n + 1 = 1$; that is, $n = 0$. In short, $a = 2$.

In the remaining possibilities we see that any prime other than 2 is of the form either $4n + 1$ or $4n + 3$. Since there is an infinite number of primes different from 2, and these must be of the form $4n + 1$ or $4n + 3$, there is an infinite number of primes of the form $4n + 1$ or an infinite number of the form $4n + 3$. It is not a priori obvious that both need be true, namely, that there be an infinity of the type $4n + 1$ *and* an infinity of the type $4n + 3$. The fact is that both are true; that is, as n varies we find an infinite number of primes in the set $\{4n + 1\}$ and in the set $\{4n + 3\}$. We shall now adapt Euclid's proof to show the existence of an infinity of primes of the form $4n + 3$. The result for $4n + 1$ is much more difficult and we leave it for later.

Before going into the details of the argument we make two remarks. The first of these is a result that we shall prove in Section 4 but which we shall use immediately. (It is a slight strengthening of the fact, proved above, that any integer greater than 1 is divisible by some prime.) The second we prove here.

1. Any integer greater than 1 is a product of prime numbers.
2. A product of integers of the form $4n + 1$ is again of the form $4n + 1$.

We prove remark 2 for the product of two integers; to extend this to a product of more than two integers repeat the process. Suppose, then, that $a = 4u + 1$, $b = 4v + 1$; then

$$ab = (4u + 1)(4v + 1) = 16\, uv + 4(u + v) + 1 = 4w + 1$$

where $w = 4uv + u + v$.

We return to the question of primes of the form $4n + 3$. Suppose that there is only a finite number of them:

$$3 = q_1 < q_2 < \cdots < q_k$$

Consider $a = q_1 q_2 \cdots q_k$ and $b = 4a - 1$. Now if $q_i | b$, then since $q_i | a$ (and hence $q_i | 4a$) we would get that $q_i | 1$, which it does not. In short, $b = 4a - 1$ is not divisible by any prime of the form $4n + 3$. Since $b = 4a - 1$, b is not divisible by 2. Therefore every prime that divides b must be of the form $4n + 1$. By remark 1 we then get that b is a product of primes of the form $4n + 1$. By remark 2, we get that b, itself, must be of the form $4n + 1$. But this is false, for b is of the form $4n - 1$. In other words, the assumption that there is a finite number of primes of the form $4n + 3$ led us to a contradiction. As a result we have that there is an infinite number of primes of the form $4n + 3$.

One can imitate the argument just given to prove that we can find an infinite number of primes of the form $6n + 5$ (Exercise 4). Will this kind of argument yield the existence of an infinity of primes of the form $8n + 7$? It does not! What is peculiar about 4 and 6 that makes the argument work for them but fail to work for 8?

It seems natural to generalize these results. A remarkable generalization of this observation is the deep and very beautiful result of Dirichlet. The result, quite difficult to prove, is easy to enunciate.

P. Dirichlet (1805–1859) was a German mathematician who did deep and fundamental work in number theory. He was married to Rebecca Mendelssohn, the composer's younger sister.

Let a, b be two integers having no common factor other than 1; that is, if $c > 0$ and $c | a$, $c | b$ then $c = 1$. Then the set of all integers $an + b$ contains an infinite number of primes.

Another way of saying this is that the arithmetic progression $an + b$ represents an infinite number of primes. Our previous result for $4n + 3$ is the special case of the Dirichlet situation where $a = 4$ and $b = 3$.

The result of Euclid and that of Dirichlet tell us that there is an infinite number of primes. But how much sharper and more incisive is that of Dirichlet! Never mind looking for primes in the whole domain of integers; it is enough to look in certain slices of the integers—in nice arithmetic progressions—and there an infinity of primes already can be found.

One immediately asks: Is the Dirichlet theorem itself but a special case of a wider phenomenon? Can we pass from arithmetic progressions to a more general family of expressions? The intuitive feeling is certainly yes, that one should be able to assert that a nicely enough defined function will yield an infinite number of primes. However at this stage of mathematical history, we are stymied in this. For instance, something so easy-sounding as: Is there an infinity of primes of the form $n^2 + 1$? is an open question, and it seems probable that it will remain open for some time yet. Here are two more innocent-looking questions whose solution today is still not available, and which seem out of reach of our methods for attacking them:

1. If we consider the integers of the form $p + 2$, where p is restricted to range over the prime numbers, do we pick up an infinity of primes? This is the famous "twin-prime" problem. Some information is known, namely, that such twin primes are not too thickly distributed among all primes; but as to their infinity, nothing.
2. If we consider the integers $2p + 1$, where p ranges over the prime numbers, do we pick up an infinity of primes? Here, too, virtually nothing is known.

The powerful methods we have for handling the Dirichlet theorem shed no light on the questions just raised. However one can produce an explicit "expression" that takes on only prime values and produces an infinite set of primes. This was first done by a young American mathematician, Mills, in 1947: If x is a positive real number let $[x]$ be the largest integer less than or equal to x. For instance, $[3.795] = 3$, $[5.1] = 5$, $[\pi] = 3$. What Mills showed was: There exists a real number $a > 1$ such that $[a^{3^n}]$ is a prime number for every positive n. Very little is known about such possible a's; after all, Mills did not give one explicit real number a, he proved its *existence*. For instance, it is not even known if such an a can be a rational number.

Later, other prime-representing functions of this kind were produced. To cite an example, one given by Sierpinski in 1952 actually represents the nth prime in terms of the function $[x]$ and a quite explicit real number a.*

One can consider other types of expressions to see how effective they are in yielding primes. For instance, let us look at the expression $n^2 - n$

---

*There is a survey of prime-representing functions by U. Dudley in *American Mathematical Monthly* 78 (1969): 23–28.

+ 11; it cannot take on *only* prime values as n takes on the values 0, 1, 2, . . . , 11, 12, ... This is seen clearly because when $n = 11$, its value is $11^2 - 11 + 11 = 121$, which is not prime. The expression is strange in one respect, however; if we tabulate $n^2 - n + 11$ for the values $n = 0, 1, 2, . . . , 10$, we see the table:

n	$n^2 - n + 11$
0	11
1	11
2	13
3	17
4	23
5	31
6	41
7	53
8	67
9	83
10	101

Notice that every value in the second column is a prime number. Nor is 11 the only such number. For lack of a better word, we call a positive number m "lucky" if the expression $n^2 - n + m$ takes on prime values for $n = 0, 1, 2, . . . , m - 1$. For m to be lucky, it must be a prime number itself (Exercise 6). As we saw above, 11 is a lucky number. A direct check verifies that the following are lucky numbers: 2, 3, 5, 11, 17, 41—six in all.

In 1934, H. Heilbronn and E. N. Linfoot proved that there could be *at most one* more lucky number. Computations revealed that if such a further lucky number existed it would have to be very large. Recently, in 1967, H. M. Stark proved that indeed there exist no other lucky numbers, so the list given above accounts for all.

In the preceding discussion we saw that there exist polynomials (such as $x^2 - x + 41$) that take on a lot of prime values. If we use the symbol f for a polynomial in x, we can rephrase this by saying that $f(x)$ may be a prime number for many integral values of x. But we did not see a polynomial f such that $f(x)$ is prime for *every* integer x. There is a very good reason for this: There is no such polynomial. This is not hard to prove, but we shall not enter into the details. The case of a polynomial of degree two and highest coefficient one appears below in Exercise 9.

The same thing is true for polynomials in more than one variable. A typical polynomial in three variables, for instance, looks like

$$2x^3 - 5x^2yz + 4xyz^7 - y^{11} + 13$$

and we are asserting that it is impossible for this expression (or any such expression) to yield nothing but prime numbers when we put in specific integers for x, y, and z. (Can you pick x, y, z so as to make this particular polynomial composite?)

But if we change the project just a little, something quite remarkable happens. In 1970, Y. B. Matijasevich, completing work begun by Martin Davis, Hilary Putnam, and Julia Robinson, proved that there exists a polynomial g such that every *positive* value g takes on is a prime and we get all the primes this way. (No doubt g takes on lots of negative values, but we simply pay no attention to these.) Matijasevich's methods are constructive and can be used to find such a polynomial explicitly; possibly this will be done before too long.

The results touched on so far have told us something about the way in which primes are distributed. We want more information about this. Can we find large "deserts" in the positive integers where there are no primes? Let us, for instance, search for two consecutive odd composite integers. The first example is given by 25, 27. We can surround these integers by even numbers so as to get the string 24, 25, 26, 27, 28 of five consecutive composite integers. The first instance of three consecutive odd composite integers is: 91, 93, 95. By adjoining the nearby even numbers we can expand this to 90, 91, 92, 93, 94, 95, 96, a sequence of seven consecutive composites. Now we make the challenge more severe. Given a positive integer n, produce n successive positive integers none of which is a prime. In other words, find an integer a (depending on n) such that none of $a, a + 1, \ldots, a + n - 1$ is a prime number. Here trial and error will not work; we need something systematic.

In Chapter 1, Section 3, we defined the symbol $m!$, for a positive integer m, by $m! = 1 \cdot 2 \cdot 3 \cdot \ldots \cdot m$. Given the positive integer n, let us try $a = (n + 1)! + 2$. Since $2|(n + 1)!$ and $2|2$, $2|a$ and so a is not a prime. Now $a + 1 = (n + 1)! + 3$; thus since $3|(n + 1)!$ and $3|3$, $3|(a + 1)$ so $a + 1$ is not a prime. In fact we can show that $i|(a + i)$ for $1 \leq i \leq n - 1$; so we see in this way that none of the integers $a, a + 1, \ldots, a + n - 1$ is a prime. Thus we can produce large stretches of integers in which no prime appears. In fact one can do even better. Although it is more difficult, one can isolate primes as a sort of oasis in a long desert of integers. Precisely, given any positive integer n, we can find a prime number p such that the n integers preceding p and the n integers succeeding it are not primes.

We give a proof of this fact. In doing so, however, we make use of the theorem of Dirichlet which states that if a and b are integers having only 1 as a common factor, then the arithmetic progression $am + b$ takes on an infinite number of values which are prime numbers.

So our task is the following: Given a positive integer $n > 1$ we want to find a prime p such that the n integers preceding p and the n integers following it are all *composite*. Another way of saying this is that we seek a prime p such that none of $p \pm 1, p \pm 2, \ldots, p \pm n$ is a prime number.

Let q be a prime number such that $q > n + 1$; we already know that such a q exists because there is an infinite number of primes. Let

$$a = (q - 1)(q - 2) \cdots (q - n)(q + 1)(q + 2) \cdots (q + n)$$

Since $n < q$ it is clear that $q \nmid a$, for otherwise q divides one of the factors of a (see Theorem 3.4 in Section 4). But for $0 < i < q$ we have both $q \nmid (q - i)$ and $q \nmid (q + i)$. Hence q is relatively prime to a. Therefore, by Dirichlet's theorem, there is an infinity of primes of the form $am + q$. Pick p to be such a prime with $m > 0$. We claim that p does the trick. Since $p = am + q$, we have $p \pm i = am + q \pm i$. Now for $0 < i < n$, $q \pm i$ is a factor of a by the very form of a. Moreover $q \pm i \geqq 2$, so it is a genuine factor of a. Since $q \pm i$ is a divisor of $q \pm i$ and of a, it is a divisor of $am + q \pm i$, hence of $p \pm i$. Thus all the integers $p \pm i$ with $i = 1$, $2, \ldots, n$ are composite and we have produced the desired prime. (The result and argument we just gave are due to the Polish mathematician Sierpinski.)

We return to our number $a = (n + 1)! + 2$; even with n small, it is relatively large. One asks: Can one do much better than a as the beginning of a prime-free stretch of n integers? For instance, for $n = 5$ we already saw that 24, 25, 26, 27, 28 work. But a here is $(5 + 1)! + 2 = 722$, which is much larger than 24. In general it seems reasonable that a much smaller number than a should do; but here very little is known. One would guess that a considerably smaller number could be found that would do the trick. However one cannot expect to drop down too far; as we shall shortly see, to get n successive nonprime integers we have to start at least at n (see the Bertrand postulate on page 42).

The purpose of the above discussion has been to show that the primes seem to be dotted among the integers in a strange way, with large empty tracts in which no prime appears. One might therefore expect that there is relatively little that can be said about the distribution of the primes. Strangely enough, this is not so. One of the very beautiful theorems in mathematics, the *prime number theorem*, tells us with a great degree of accuracy how many primes there are less than a given integer. The result, while not recent, is not very old. It was conjectured by Gauss about 1840, who arrived at his conjecture by inspecting a table of primes; this was proved independently in 1896 by de la Vallée-Poussin and Hadamard, using powerful weaponry from areas of mathematics far removed from number theory. More recently, in 1946, an elementary but not easy proof was given it by Erdös and Selberg. We now go on to describe this result.

One can hardly begin to describe all that C. F. Gauss (1777–1855) contributed to mathematics. By general consensus, this German mathematician, together with Archimedes and Newton, is considered one of the greatest mathematicians that ever lived. His genius was already clear at the early age of 6, and his mathematical powers did not abate until his death. Everything he touched and created became important. Outside of mathematics, he is famous for work he did in physics and in astronomy. There is a sketch of his life in Bell's *Men of Mathematics,* and in addition two full length biographies:

Guy W. Dunnington, *Carl Friedrich Gauss, Titan of Science,* Exposition Press, 1955.

Tord Hall, *Carl Friedrich Gauss, a Biography,* translated by Albert Froderberg, Cambridge, Mass.: MIT Press, 1970.

J. Hadamard (1865–1963) was a great French mathematician whose work in the thory of functions (much of this early work was designed to attack the prime number theorem) is among the classic work in this area. On the personal side, he was a brother-in-law of Dreyfus, whose trial on trumped-up charges of espionage led to a world outcry against French military "justice" at the end of the nineteenth century.

C. J. de la Vallée-Poussin (1866–1963) was a noted Belgian mathematician.

If n is a positive integer let $\pi(n)$ denote the number of primes less than or equal to n. Thus, for instance, $\pi(5) = 3$ since 2, 3, 5 are the primes less than or equal to 5, $\pi(15) = 6$, and so on. We want to know roughly how large $\pi(n)$ is; we also want to know the precise meaning of this "roughly." This is the content of the prime number theorem. There is a certain mathematical function, from the positive real numbers to the real numbers, which one can calculate and tabulate, called log x, the logarithm of x to the base e. Here e is a fundamental mathematical constant, given to nine decimal places by $e = 2.718281828. \ldots$ The prime number theorem then asserts that $\pi(n)$ is roughly equal to $n/\log n$ for large n. More precisely, it says that the ratio $\pi(n)/(n/\log n)$ hovers as close to 1 as we like if n is large enough. Even more precisely it says: Given any number $\varepsilon > 0$, say $\varepsilon = .001$, we can find an integer n_o (which depends on ε) large enough so that for all $n > n_o$, $\pi(n)/(n/\log n)$ falls between $1 - \varepsilon$ and $1 + \varepsilon$ (so, for the particular case $\varepsilon = .001$, $.999 < \pi(n)/(n/\log n) < 1.001$).

The theorem is a kind of averaging result; it ignores the gaps and overabundances of primes in regions of the integers, and smooths out everything to give rather precise information about not "how," but "how often," the primes occur in the integers. As a consequence of the prime number theorem we get a good approximation of the mth prime. If we enumerate the primes as

$$p_1 = 2 < p_2 = 3 < p_3 = 5 < \cdots < p_m < \cdots$$

and call p_m the mth prime, then it follows from the prime number theorem that p_m is of the order of magnitude $m \log m$.

Some years prior to the important work of de la Vallée-Poussin and Hadamard, the mathematician Chebycheff in 1850 established some remarkable properties of the function $\pi(n)$. He proved that for all $n > 1$

$$\frac{7}{8} \frac{n}{\log n} \leq \pi(n) \leq \frac{9}{8} \frac{n}{\log n}$$

While for large n this is not as sharp as the prime number theorem, it has the virtue of holding for all $n > 1$.

P. Chebycheff (1821–1894) was one of the leading Russian mathematicians of the nineteenth century.

From this he deduced a very lovely result which asserts that for every positive integer n there is a prime number between n and $2n$. This is often called the Bertrand postulate (the name is misleading because it is a theorem). Nor is the number 2 sacred in this. For instance, if we are willing to settle for a statement that is valid for *large n*, we can replace 2 by $\frac{4}{3}$ or $\frac{6}{5}$ or indeed any number larger than 1. The number 2 is special in that between n and $2n$ there is *always* a prime; that is, we do not have to wait until n is large enough.

Joseph Louis François Bertrand (1822–1900) was a French mathematician who made contributions to analysis, thermodynamics, and probability theory. It was in 1845 that he made the conjecture named after him and verified it up to 6,000,000.

Since we have been discussing the Bertrand postulate it seems natural to bring up questions that have the same flavor. For instance, is there a prime between n^2 and $(n + 1)^2$ for every positive integer n? This question is open. Little is known about this very specific question. However, if we change it slightly we can get definitive results. To cite one, it has been proved that there is always a prime between n^3 and $(n + 1)^3$. This fact played a vital role in the work of Mills, described earlier, constructing a function whose value, at every positive integer, is a prime number.

In what has been said we see a nice example of successive deepening of a mathematical result. Euclid's theorem merely asserts that an infinity of prime numbers exists, Dirichlet's result says more, that such an infinity exists in selected parts of the integers. The Bertrand postulate deepens Euclid in another direction, namely, that primes appear not too infrequently, some prime cropping up between any positive integer and its double. Finally, the prime number theorem counts in a highly accurate fashion how many primes there are up to a given point, and predicts how large, in order of magnitude, the nth prime must be.

Still another famous old open question about prime numbers is known as the *Goldbach conjecture*. It asks whether every even integer larger than 4 is the sum of two odd primes. The very nature of the problem is such that it is bound to be difficult. After all, the primes were defined in terms of the *multiplicative* properties of the integers, whereas the statement of the conjecture is about the *additive* nature of the primes. Although the Goldbach conjecture is not solved, very much is known about it. In a very profound, difficult, and brilliant work, Vinogradov, in 1937, showed that every large enough odd number is the sum of three primes.

In closing this section we shall mention some recent computations connected with Euclid's proof that there are infinitely many primes. They appear in the April, 1972 issue of the journal *Mathematics of Computation* (vol. 26, pp. 567–570) and are due to Alan Borning, in his undergraduate thesis at Reed College.

Let p be a prime. The reader may have wondered: Is the expression $2 \cdot 3 \cdot 5 \cdots \cdot p + 1$ used by Euclid to get a new prime already a prime itself? Here is the answer for $p \leq 307$: It doesn't happen very often, in fact, only for $p = 2, 3, 5, 7, 11$, and 31.

The expression $2 \cdot 3 \cdot 5 \cdots \cdot p - 1$ could equally well have been used by Euclid. It too is prime just six times up to $p = 307$, for the values $p = 3, 5, 11, 13, 41$, and 89.

From Borning's paper we cite also the large numbers $27! + 1$ and $30! - 1$, which he found to be prime.

Exercises

1. Show that, to check whether or not a given positive integer n is a prime, we must merely check whether n is divisible by some prime less than or equal to $\sqrt{n}$.
2. Use the result of Exercise 1 to prove that 421 is a prime.
3. Prove the following properties of divisibility:
   (a) if $a|b$ and $b|c$, then $a|c$
   (b) if $a|b$ and $b|a$, then $a = b$ or $a = -b$
4. By modifying Euclid's method, prove that there is an infinite number of primes of the form $6n + 5$.
5. Find all prime numbers less than 200.
6. Prove that if m is a lucky number (see page 38), then it must be a prime number.
7. Verify that 17 is a lucky number.
8. Verify the Goldbach conjecture for all even numbers between 6 and 50.
9. If a and b are integers, prove that $n^2 + an + b$ cannot take on prime values for every value of n.
10. If p is a lucky number, prove that $n^2 + n + p$ takes on prime values for $n = 0, 1, 2, \ldots, p - 2$.
11. Let $m > 1$ be an integer; prove that exactly one integer in any chain of m consecutive integers is divisible by m.
12. Prove that $n(n + 1)(n + 2)/6$ is an integer for any integer n.
13. Prove that $n(n + 1)(n + 2)(n + 3)/24$ is an integer for any n.
14. Show that there are infinitely many primes p such that $p - 2$ is not a prime. (Hint: Use Exercise 4.)
15. (a) A "prime triple" might be defined as a succession of three primes of the form $p, p + 2, p + 4$. Example: 3, 5, 7. Prove that this is the only prime triple.

(b) A prime quadruple is a succession of four primes of the form $p, p + 2, p + 6, p + 8$ (note the gap in the middle). The first prime quadruple is 5, 7, 11, 13. Show that in any larger prime quadruple $p + 4$ must be divisible by 15. Find the next two prime quadruples. (Optimists conjecture that prime quadruples go on forever, and even more far-reaching conjectures have been made.)

(c) Go back to part (a) and give an improved definition of "prime triple." You should in fact define two kinds of prime triples. Give the first three examples of each.

16. (The "6174 game") Take any four-digit number, assuming only that the digits are not all the same. We illustrate with the number 4527. Write the digits in descending order and in ascending order: 7542 and 2457. Subtract: $7542 - 2457 = 5085$. Repeat the operation: $8550 - 0558 = 7992$. And so on: $9972 - 2799 = 7173$; $7731 - 1377 = 6354$; $6543 - 3456 = 3087$; $8730 - 0378 = 8352$; $8532 - 2358 = 6174$; $7641 - 1467 = 6174$, and of course from now on there will be nothing new.

Try this for a few examples yourself. (Remark: Numbers with less than four digits are fine; put 0's in front. For instance, starting with 427 we get $7420 - 0247 = 7173$, etc.)

It is a fact that after a few steps you always reach 6174. With good organization of the work, you can check this in about a half hour of hand work. Or, if you are handy at programming and have access to a computer, do it that way. The best idea is to work up a theory that avoids any serious computation.

With some accompanying patter and a receptive audience, you can make a successful magic trick out of the 6174 game.

17. Is there something like the 6174 game for two-digit or three-digit numbers?

18. (The "Syracuse algorithm") Here is a somewhat fancy definition of a function F, defined on the set S of all positive odd integers, and taking values in S. If x is a positive odd integer, let $F(x)$ be the result of dividing $3x + 1$ by the highest power of 2 which divides it. For instance, if $x = 7$, then $3x + 1 = 22$, the highest power of 2 dividing 22 is 2, and so $F(7) = 11$. Again, if $x = 9$, we have $3x + 1 = 28$, and $F(9) = 28/4 = 7$.

(a) Make a table of $F(x)$ for $x = 1, 3, 5, \ldots, 19, 21$.

(b) Is F one-to-one?

(c) Is F onto? (Hint: Can $F(x)$ be divisible by 3?)

(d) Now try the result of applying F repeatedly. For instance, $F(11) = 17$, $F(17) = 13$, $F(13) = 5$, $F(5) = 1$. Check for all other values of x up to $x = 25$ that on applying F repeatedly to x you eventually reach 1.

(e) (Try this only if you are patient.) Apply F repeatedly to 27 until you reach 1. (If all goes well you should encounter 42 numbers in all, counting both 27 and 1.) What is the biggest number that you encounter?

(It is not known whether the Syracuse algorithm must always end in 1, starting with any odd number. This has been checked extensively by computer.)

19. Define a modified "Syracuse algorithm" G by using $3x - 1$ instead of $3x + 1$. Check the behavior of small odd numbers under repeated application of G. (You should find that there is quite a difference, by locating two numbers x and y that run in a circle: $G(x) = y$, $G(y) = x$.)

20. (a) Recall (page 38) that a prime p is *lucky* if the numbers $n^2 - n + p$ are prime for all the values of n from 0 to $p - 1$, and that it is hopeless to continue to $n = p$. Show that $n = p + 1$ is likewise hopeless, by simplifying and factoring $(p + 1)^2 - (p + 1) + p$.

(b) It turns out that the next value ($n = p + 2$) is *not* hopeless. In fact, let us call a lucky prime p "super-lucky" if $n^2 - n + p$ yields a prime for $n = p + 2$. Check that of the six lucky primes (2, 3, 5, 11, 17, 41) all are super-lucky except the first. (A small table of primes would be handy for this question and for the next one.)

(c) Make a similar investigation for $n = p + 3$.

(d) Show that $n^2 - n + p$ factors algebraically for $n = p + 4$.

21. In this exercise we propose a different kind of lucky number. Let p be an odd number. Call p an L-number if $n^2 + p$ is prime for $n = 0, 2, 4, \ldots, 2p - 2$. Find two L-numbers. (This is believed to be a new question. Finding all L-numbers will presumably be as hard as finding all lucky numbers.)

2. Some Formal Aspects

We just finished speaking rather informally about the integers. However, in all this talk, we never lost sight of the fact that the object under discussion was an integer, and hence enjoyed some special properties. We now turn in a more formal direction, one in which particular integers do not play much of a role, but where, instead, the whole set is considered relative to combination by addition and multiplication.

What we will be doing is really going through the beginning steps of the theory of rings—a part of abstract algebra. We will do so in the next three sections. Eventually we will come back to earth, to the good old properties of integers, primes, squares, and the like. In the meantime we become more formal.

Given two integers, we can combine them in several ways with an outcome that is again an integer. For instance, we can add them, subtract them, or multiply them together, resulting always in another integer. Division is something else; here we have no guarantee that the end product of our calculation will end up in what we have selected, perhaps arbitrarily, as the desired set. For example 6 and 7 are highly respectable integers, yet $\frac{7}{6}$ fails to be an integer.

Let us take the point of view that above all, we want our operations on the objects at hand to carry us back into the set with which we started. In this way we may at least limit the possible range of objects under discussion. Instead of exhibiting an interest in each object, per se, as an object, suppose we concentrate on the interplay among these objects in the whole community of integers.

All the properties of the integers under addition and multiplication can be firmly established starting with a rather modest set of assumptions or axioms. However let that not be our concern; suppose that we are all familiar with the usual properties of integers under + (addition) and · (multiplication). We list some of these below. One may ask: Why pick these particular properties in the host of possible choices? A perfectly reasonable question! The answer is rather smug and dogmatic: With hindsight one has come to realize that these, at least formally, play the basic role.

For convenience, let Z denote the set of integers. The basic properties we list are as follows:

1. If $a, b \in Z$, then $a + b \in Z$. We describe this by saying that Z is *closed* under *addition*.
2. If $a, b, c \in Z$, then $(a + b) + c = a + (b + c)$. This is the so-called *associative law* of *addition*.
3. If $a, b \in Z$, then $a + b = b + a$. This is the *commutative law* of *addition*.
4. There is a particular integer, which we write as 0, such that $a + 0 = a$ for every integer $a \in Z$.
5. Given any integer $a \in Z$ we can find a suitable $b \in Z$ such that $a + b = 0$. We shall later write b as $-a$.

These five postulates about the addition are all that we shall require concerning the additive nature of Z. Note that these requirements are *formal*, the *meaning* of the integers plays no part. We now turn to the multiplicative nature of Z. We write ab for $a \cdot b$.

The formal rules satisfied by the product are outlined in:

(i) If $a, b \in Z$, then $ab \in Z$; Z is said to be *closed* under *multiplication*.

(ii) If $a, b, c \; \varepsilon \; Z$, then $(ab)c = a(bc)$; this is the *associative law* of *multiplication*.

(iii) If $a, b \; \varepsilon \; Z$, then $ab = ba$; this is the *commutative law* of *multiplication*.

(iv) There is a special integer, which we write as 1, such that $a1 = a$ for all $a \; \varepsilon \; Z$.

Note the analogy between the rules for the addition and those for the multiplication. Note that 0 is to addition as 1 is to multiplication; that is, under one of the operations 0 always restores an element to itself, under the other 1 does the same thing. Note, too, the lack of a (v) to match the (5); it would be: Given $a \; \varepsilon \; Z$ there is $b \; \varepsilon \; Z$ with $ab = 1$. This is false in Z; ignoring the special behavior of 0, even a decent integer such as 2, for instance, lacks an *integer mate b* with $2b = 1$.

So far, in (1) through (5) and in (i) through (iv) we have described the behavior of the sum and product, respectively, by themselves; as yet we have not interlocked them. But these operations do not live in ignorance of each other. They are tied together by

(D) $(a + b)c = ac + bc$ for all $a, b, c \; \varepsilon \; Z$

This is the *distributive law*.

These above-delineated rules, or facts, or what-you-will, of a formal nature are the only ones we shall set down. At the risk of boring the reader with "obviosities" we repeat again: These are not God-given, holy rules, nor empirical rules—they follow, not from thin air, but from assumptions made in set theory in defining the concept of integer. There is no question or desire to establish them as absolutes founded on no assumptions. Nor do we take cognizance in them of what an integer is, merely how it behaves.

Also we should recognize that the integers are something special, for in them the product is determined by the sum. What does this mean? If we restore its meaning to the integer 2, then $2a$ comes out as $a + a$; similarly, we can interpret ab in terms of sums (even when a or b is negative). So the additive and multiplicative pieces described above are not independent. But let us behave as if they were.

Now one asks, do those carefully written assumptions, ten in number, characterize the integers? Again, what does this mean? In our description we ignored the empirical meaning, or the cardinal numeric meaning, of $+$ and $\cdot$; they are merely formal operations on some gadgets. We claim that we can construct other formal objects with a formal addition and multiplication that satisfy all our requirements. Here is a simple example.

Let A be a set consisting of the two objects □ and *. In A we *define* an addition which we denote as $+$ by spelling out for it all the possibilities. In fact we decree that:

$$\square + \square = \square, \square + * = *, * + \square = *, * + * = \square$$

We now *define* a multiplication which we denote as · by describing what happens explicitly in

$$\square \cdot * = \square, * \cdot \square = \square, * \cdot * = *, \square \cdot \square = \square$$

It is routine, albeit laborious, to verify that A and the operations $+$ and $\cdot$ introduced in A satisfy all the formal requirements laid out in (1), . . . , (5), (i), . . . , (iv), (D). We do a sample verification. Our requirement (D) demands that given a, b, c then $(a + b) \cdot c = a \cdot c + b \cdot c$. Let us try this with $a = \square$, $b = *$, $c = *$. Then we would like $(\square + *) \cdot *$ to equal $\square \cdot * + * \cdot *$. Now, by our very definition of $+$, $\square + * = *$ so that $(\square + *) \cdot * = * \cdot *$; returning to our definition of $\cdot$ we see that $* \cdot * = *$. Hence $(\square + *) \cdot * = *$. On the other hand, from the definition of $\cdot$, $\square \cdot * = \square$ and $* \cdot * = *$ so that $\square \cdot * + * \cdot * = \square + * = *$. Thus we see that $(\square + *) \cdot * = \square \cdot * + * \cdot *$, for each side of this relation worked out to be equal to $*$.

The $\square$ in A *acts* as the required 0 should, the $*$ as the required 1. Having produced a model, A, which formally, at least, acts like the integers under $+$ and $\cdot$ and is certainly not the set of integers, we conclude that our formal setup is not stringent enough to pin down and single out the set of integers.

It is this type of speculation that often gives rise to abstraction in mathematics. All right, so the integers are not the only systems described by our axioms or rules! Can we at least establish something worthwhile about *all* systems so described? Whatever holds true in such a general setting will certainly hold true for any special model satisfying these conditions, therefore, will hold for the set of integers.

Now comes a problem: To what extent should we abstract? We can subject a system to satisfy (1) through (5); such objects are called *abelian groups*. In other words, an abelian group G is a nonempty set of elements in which is defined a single operation, say $+$, satisfying:

(1) $a, b \; \varepsilon \; G$ implies $a + b \; \varepsilon \; G$.
(2) $a, b, c \; \varepsilon \; G$ implies $(a + b) + c = a + (b + c)$.
(3) $a, b \; \varepsilon \; G$ implies $a + b = b + a$.
(4) There is a particular element, which we denote by 0, such that $a + 0 = a$ for all $a \; \varepsilon \; G$.
(5) Given $a \; \varepsilon \; G$ we can find an element $b \; \varepsilon \; G$ such that $a + b = 0$.

We can push closer to the integers and study systems satisfying all the rules we set down, that is, systems with two operations which we choose to denote by $+$ and $\cdot$, and which satisfy all the rules (1), . . . , (5), (i), . . . , (iv), (D). Such mathematical systems are called *commutative rings* with unit element.

Exercises

1. In the example $A = \{\square, *\}$ discussed, verify:
   (a) $(\square \cdot \square) \cdot * = \square \cdot (\square \cdot *)$
   (b) $(\square + \square) \cdot * = \square \cdot * + \square \cdot *$
   (c) $(\square + *) + \square = \square + (* + \square)$
   (d) $(* + *) \cdot \square = * \cdot \square + * \cdot \square$

2. Let S be a set and let $\mathcal{B}$ be the set of all subsets of S. For $x, y \in \mathcal{B}$ (i.e., x and y are subsets of S) define $x + y = x \cup y$, the union of x and y. Verify which properties in (1) through (5) hold in A with $+$ so defined. What element of $\mathcal{B}$ (i.e., what subset of S) acts as 0 should?

3. Let S be a set and let $\mathcal{B}$ be the set of all subsets of S. For $x, y \in \mathcal{B}$ define $x \cdot y = x \cap y$, the intersection of x and y. Verify that (i) through (iv) hold in $\mathcal{B}$ for this multiplication, $\cdot$, so defined. What element of $\mathcal{B}$ (i.e., what subset of S) acts as 1 should?

4. For $\mathcal{B}$, $+$ and $\cdot$ as in Exercises 2 and 3 verify that (D) holds, namely, that $(x + y) \cdot z = x \cdot z + y \cdot z$. (Hint: See Exercise 20 in Section 2 of Chapter 1.)

5. Let N be the set of positive integers. For $a, b \in N$ define $a \circ b = a^b$. Is $a \circ b = b \circ a$?

6. Let E be the set of all even integers with the usual addition and multiplication. Is E a commutative ring with unit? If not, which of (1), $\ldots$, (5), (i), $\ldots$, (iv), (D) fail to hold?

3. Some Formal Consequences

We remain, for a little while longer, on the formal level. We want to derive some arithmetic consequences from our axioms, and we want to do so in a logically sound way. Because we are operating not really in the integers but in any commutative ring with unit, the results so obtained hold in this wider setting. We are not saying that these results are of deep intrinsic interest; for the integers they certainly are not, and even for the more general model they could hardly be described as exciting. But what do they do? For one thing they illustrate what a proof is, even if at a low level; they show a connected, consistent, logical deduction from a system of postulates to a result holding in this system. Even more than this, the arithmetic axioms may clear up some doubts or misunderstandings about elementary things which, during our school days, have too often been wrapped up in some vague obscurity and held up to us as deep or self-evident facts about numbers. The truth is that they are neither deep nor self-evident and they do not have much to do with numbers. They are consequences of some formal structure, embodied in our postulates for the addition and multiplication, without any deep or hidden significance.

Let us consider one such result, the one that says that $(-1)(-1) = 1$. What arcane mystique has been attached to this! What ridiculous "proofs" and justifications by analogy have been given it! Double negatives—in English at least—mean the positive; true, but what has this to do with $(-1)(-1)$? It is interesting to read what a highly intelligent human being as Stendhal—who once aspired to be a mathematician if only to escape from Grenoble—writes about the various attempts of his teachers and fellow students to convince him of the truth of $(-1)(-1) = 1$. No wonder he was never convinced! We quote from his charming, gossipy autobiography *Henri Brulard*.

> In my opinion, hypocrisy was impossible in mathematics, and in my youthful simplicity I thought it was also the case in all sciences to which I had heard they were applied. What were not my feelings when I perceived that nobody could explain to me how it came about that minus multiplied by minus gives plus $(- \times - = +)$? (This is one of the fundamental bases of the science called *algebra*.)
>
> They did worse than fail to explain this difficulty (which is doubtless explicable, for it leads to truth): they explained it to me by reasoning which was obviously not very clear to those who gave it.
>
> M. Chabert, when I pressed him, got confused, repeated his *lesson,* the identical one against which I was advancing objections, and ended by saying to me, in effect:
>
> 'But it is the custom; everybody admits this explanation. Euler and Lagrange, who were presumably as good as you, admitted it.'
>
> · · ·
>
> I distinctly remember that when I spoke of my difficulty about 'minus multiplied by minus' to one of the clever ones, he would laugh in my face; they were all more or less like Paul-Emile Teisseire, and used to learn by heart. I often heard them say, at the blackboard, at the end of their demonstration:
>
> 'And so it is evident that,' etc.
>
> 'Nothing is less evident to you,' I used to think. But for my part, I was dealing with things that were evident, and which, with the best will in the world, it was impossible to doubt.
>
> Mathematics only deals with a small corner of things (their quantity), but on this point it has the advantage of saying nothing but what is certain, the truth, and almost the whole truth.
>
> At the age of fourteen, in 1797, I used to imagine that the higher mathematics, those that I had never known, dealt with *all* or almost all sides of things, and that in this way, as I proceeded, I should arrive at a knowledge of things, sure and indubitable and demonstrable at will, *about everything*.
>
> It took me a long time to convince myself that my objection about 'minus multiplied by minus gives plus' was absolutely incapable of entering the head of M. Chabert, that M. Dupuy would never answer it except by a superior smile, and that the clever boys whom I asked questions would always make fun of me.

It is interesting to read Stendhal's opinions and attitudes about certain mathematicians and mathematical personages of his day, attitudes formed by studying from their elementary textbooks. It is fortunate that his insight into human nature was keener than his evaluation of mathematicians.

It is also interesting to read about his being perplexed by something that was mysterious, on first seeing it, to all of us. How universal and timeless is human reaction and human response!

We wish to establish certain rules in Z that will allow us to make calculations as they arise. Although we are working seemingly with integers, all that we are really using are the rules (1), ... , (5), (i), ... , (iv), (D), *hence what we establish will hold in any commutative ring wtih unit element.*

We start things off with

Lemma 2.1. If $a, b, c \varepsilon Z$ and $b + a = c + a$, then $b = c$.

Proof. By property 5 of addition, there is an $x \varepsilon Z$ such that $a + x = 0$. Hence $(b + a) + x = (c + a) + x$. Using the associative law of addition we obtain $b + (a + x) = c + (a + x)$ and so, from $a + x = 0$, we see that $b + 0 = c + 0$. Since $b = b + 0$ and $c = c + 0$ the desired result $b = c$ comes out.

Note that in the proof we made no use of the multiplicative property of the integers; hence the result allowing cancellation—for this is what the lemma asserts—holds in any abelian group.

The result has some immediate consequences, consequences that also hold for any abelian group, namely the

Corollary. In Z
1. 0 is unique.
2. Given $a \varepsilon Z$ there is a *unique* $b \varepsilon Z$ such that $a + b = 0$. We write b as $-a$.
3. $-(-a) = a$ for all $a \varepsilon Z$.

Proof. Before proving the result one should explain the meaning of the word "unique" in parts 1 and 2. In part 1 it means that if Δ is an element in Z such that $a + \Delta = a$ for all $a \varepsilon Z$, then $\Delta = 0$. In part 2 we mean that if $a + b = 0$ and $a + c = 0$, then $b = c$. Now to the proofs themselves.

If $a + \Delta = a$, then we have $\Delta + a = a + \Delta = a = 0 + a$; by the cancellation property (Lemma 2.1) we conclude that $\Delta = 0$. If $a + b = 0$ and $a + c = 0$, then

$$0 = a + b = b + a = a + c = c + a$$

Canceling the a, as we are allowed to do in light of Lemma 2.1, from

$b + a = c + a$ we get $b = c$. We write $-a$ for this b; that is, $-a$ is the unique element in Z such that $a + (-a) = 0$.

By definition, $(-a) + (-(-a)) = 0$; however $0 = a + (-a) = (-a) + a$. Since this says that $(-a) + a = (-a) + (-(-a))$ we derive $a = -(-a)$.

In the lemma just proved, and in its corollary, we ignored the multiplication in Z and just considered it as a unioperational system, in fact as an abelian group under $+$. We now take cognizance of the existence of the multiplication in Z and prove

Lemma 2.2.   In Z

1. $a0 = 0a = 0$ for all $a \, \varepsilon \, Z$.
2. $a(-1) = -a$ for all $a \, \varepsilon \, Z$.
3. $(-1)(-1) = 1$

Proof.   From the properties of 0 expounded in axiom 4, $0 + 0 = 0$, hence $a(0 + 0) = a0$. Using the distributive law (D), $a(0 + 0) = a0 + a0$. Hence $a0 + a0 = a0 = 0 + a0$. By the cancellation law of Lemma 2.1, we can cancel the $a0$ from both sides leaving us with $a0 = 0$.

Now from its definition, $1 + (-1) = 0$, hence $0 = a0 = a(1 + (-1))$. Expanding by means of the distributive law gives us $a1 + a(-1) = 0$. But $a1 = a$, so that $a + a(-1) = 0$. But $a + (-a) = 0$. Then Lemma 2.1 (or its corollary) now tells us that $-a = a(-1)$. The commutativity of multiplication in Z allows us to conclude that $(-1)a = -a$.

That $(-1)(-1) = -(-1)$ follows from the general fact that $(-1)(a) = -a$ with $a = -1$. By the corollary to Lemma 2.1, $-(-1) = 1$, hence we end up with $(-1)(-1) = 1$.

Returning to our little discussion made earlier and to the plaintive account by M. Stendhal, we see that the consequence $(-1)(-1) = 1$ is *formal*, that it follows from the assumption that the "reasonable" properties (1), ..., (5), (i), ..., (iv), (D) hold for the integers, that it has nothing to do with size, debts, analogies. In fact it holds in the very general setting of a commutative ring with unit.

We are now entitled to identify $(-1)a$ with $-a$; we shall write $a + (-b)$ as $a - b$. If $x + y = z$ then $x = (x + y) - y = z - y$. This is the usual grade school rule of "transpose to the other side." This rule has been shown, in the preceding discussion, to be valid not only in the integers but in the more general framework of any abelian group and any commutative ring.

We have seen by example that the formal rules we set down to describe some of the behavior of the integers under addition and multiplication fail to single out the integers in the family of commutative rings. We shall soon try to approximate even more closely to the integers by imposing more structure or conditions on our addition and multiplication. Before doing so we consider two examples that resemble the integers in some formal aspects and which, in others, are vastly different.

Given an integer n we can divide n by 6 to get a remainder which is 0, 1, 2, 3, 4, or 5. We write $[a]$ for the remainder of a on division by 6 and let A be the set of all $[a]$. A is a set with six distinct elements, namely, $[0]$, $[1]$, $[2]$, $[3]$, $[4]$, and $[5]$. Into A we introduce two operations, which we write as $+$ and $\cdot$, by means of:

1. $[a] + [b] = [a + b]$
2. $[a] \cdot [b] = [ab]$

Note that in these operations the $+$ and $\cdot$ on the left sides are the formal ones we are introducing; on the right sides they are the ones we already have in the integers. Thus for instance, $[5] + [4] = [5 + 4] = [9] = [3]$ since 3 is the remainder of 9 on division by 6. Similarly $[5] \cdot [-4] = [5 \cdot (-4)] = [-20] = [4]$. We claim that A satisfies all the conditions for a commutative ring with unit. The element $[0]$ acts as a 0 for addition, for $[a] + [0] = [a + 0] = [a]$; the element $[1]$ acts as a 1 for multiplication, for $[a] \cdot [1] = [a \cdot 1] = [a]$. What serves as $-[a]$? Noting that we can find an element c in the range 0, 1, . . . , 5 with $[a] = [c]$ and that $[c] + [6 - c] = [c + 6 - c] = [6] = [0]$, we see how to construct the negative of an element. The verification of the formal postulates (1), . . . , (5), (i), . . . , (iv), (D) is straightforward but tedious. We shall say more about this in Section 6 of this chapter.

In many respects A is like the integers. In one important respect it is most dissimilar. Considering $[2] \cdot [3]$ we get $[2] \cdot [3] = [2 \cdot 3] = [6] = [0]$. In A it is thus possible that the product of two nonzero elements is zero, something that cannot happen in the integers.

We imitate what we did for 6 for another integer, namely, 7. As we shall shortly see the story here turns out to be different both from the integers and from the case above of 6.

Hence, let $\{a\}$ be the remainder of the integer a on division by 7 and let B be the set of all $\{a\}$. B consists of seven distinct elements, namely, $\{0\}$, $\{1\}$, . . . , $\{6\}$. We define $+$ and $\cdot$ in a similar fashion to that above by means of:

1. $\{a\} + \{b\} = \{a + b\}$
2. $\{a\} \cdot \{b\} = \{ab\}$

As before, B is a commutative ring with unit. However, here we claim that if $\{a\} \cdot \{b\} = \{0\}$, then either $\{a\} = \{0\}$ or $\{b\} = \{0\}$. Thus, in this respect, B more closely resembles the integers than does A. Moreover, we claim that given $\{a\} \neq \{0\}$ we can find a "mate" $\{b\} \neq \{0\}$ such that $\{a\} \cdot \{b\} = \{1\}$. We enumerate these:

$$\{1\} \cdot \{1\} = \{1\}, \{6\} \cdot \{6\} = \{1\}$$
$$\{2\} \cdot \{4\} = \{1\} = \{4\} \cdot \{2\}$$
$$\{3\} \cdot \{5\} = \{1\} = \{5\} \cdot \{3\}$$

In this regard B is very unlike the integers, for in the integers only 1 and -1 have such multiplicative inverses.

This last property of B makes it a highly desirable commutative ring; such systems are called *fields*. A *field* is a commutative ring F, with unit element 1, such that for every $a \neq 0$ in F there is a $b \, \varepsilon \, F$ such that $ab = 1$. The rational numbers, that is, the set of all fractions m/n, m, n integers, $n \neq 0$, under the usual addition and multiplication, form a field, the field of rational numbers. We invite the reader to verify by a direct calculation that if $\{a\} \neq \{0\}$ is in B, then $\{a\}^6 = \{a\} \cdot \{a\} \cdot \{a\} \cdot \{a\} \cdot \{a\} \cdot \{a\} = \{1\}$ (Exercise 2).

It is reasonable to wonder why the rings constructed from 6 and 7 should differ so from each other in their multiplicative behavior. The answer, as we shall see later, lies in the fact that 7 is a prime number whereas 6 is not.

One further remark is in order. There is nothing sacred about the numbers 6 and 7 that entered in defining the rings A and B respectively. We can imitate this for any positive integer n to construct, formally, a ring with n elements. This, too, we shall presently do.

We now return to the ring of integers, Z, itself. In addition to the formal properties of addition and multiplication, the integers enjoy still another structure, that of size. As we all know, the integers are divided into three groups, namely, the positive ones $1, 2, 3, \ldots$, the negative ones $-1, -2, -3, \ldots$, and the number 0. Given any integer it is one and only one of: positive, negative, or zero. We write $a > 0$ for a positive, $a < 0$ for a negative and, of course, $a = 0$ if a is zero. Note that the set of positive integers, $Z^+ = \{1, 2, \ldots\}$ enjoys:

1. If $a, b \, \varepsilon \, Z^+$, then $a + b \, \varepsilon \, Z^+$.
2. If $a, b \, \varepsilon \, Z^+$, then $ab \, \varepsilon \, Z^+$.
3. If $a \neq 0$ is in Z, then a or $-a$ is in Z^+.

However, Z^+ is not an abelian group under addition for it has no zero element or negatives for its elements.

We introduce an *ordering* into the integers from this notion of positivity. If $a, b \, \varepsilon \, Z$, we declare a *bigger* than b if $a - b > 0$. We write this as $a > b$ (or $b < a$). Note that for any two integers a and b one and only one of $a > b$, $b > a$ or $a = b$ prevails. We shall often write $a \geqq b$ (or $b \leqq a$) if $a > b$ or $a = b$.

This ordering behaves well with respect to the algebraic operations in

Z. We leave as exercises (see Exercise 4) to the reader the verification of the following:

4. If $a > b$ and $b > c$, then $a > c$.
5. If $a, b \, \varepsilon \, Z$ with $a > b$, then, for any $c \, \varepsilon \, Z$, $a + c > b + c$.
6. If $a, b \, \varepsilon \, Z$ with $a > b$, then, for $c > 0$ in Z, $ac > bc$.
7. If $c > 0$ in Z and $ac > bc$ where $a, b \, \varepsilon \, Z$ then $a > b$.

Note that 6 and 7 are false if $c < 0$, for take $a = 1$, $b = 2$, and $c = -1$; then $a = 1 < b = 2$ yet $ac = -1 > bc = -2$, for $(-1) - (-2) = -1 + 2 = 1 > 0$.

It is natural to wonder if adjoining this ordering to the already described properties of $+$ and $\cdot$ is enough to pick out the integers from the whole mass of possible commutative rings. The answer is still no.

For instance, we can give a simple, familiar ring which has an ordering and operations $+$ and $\cdot$ satisfying all the above properties, and yet the ring is not the ring of integers. Such a ring is the ring of rational numbers (fractions with numerator and denominator integers).

We present a second example that is somewhat more "exotic." In this example the properties above are satisfied, with the exception of the one asserting that the product of positive elements is positive (statement 2), and its corollary (statement 6). A similar (but fancier) example could be constructed so as to satisfy everything.

Let R be the set of all ordered pairs of integers (a, b). We declare that $(a, b) = (c, d)$ if and only if both $a = c$ and $b = d$. Addition and multiplication are introduced into R to render it a ring. The addition is defined by $(a, b) + (c, d) = (a + c, b + d)$ and the multiplication by $(a, b) \cdot (c, d) = (ac, bd)$. In other words, we operate separately on the first and second entries of the pairs. With $(0, 0)$ as the zero element, $(-a, -b)$ as the negative of (a, b), and $(1, 1)$ as the multiplicative unit element, one can verify that R is a commutative ring with unit element. Note that in R it is quite possible for the product of two nonzero elements to be the zero element. For instance, $(1, 0) \neq (0, 0)$, $(0, 1) \neq (0, 0)$ yet $(1, 0) \cdot (0, 1) = (0, 0)$.

On R we introduce an ordering, often called the *lexicographic ordering*. This name comes from the fact that we proceed as one orders words in a dictionary, according to the first letter, then according to the second, and so on. Given (a, b), (c, d) in R we define $(a, b) > (c, d)$ if:

1. $a > c$ or
2. in case $a = c$, if $b > d$

Let us see some examples of this ordering. For instance $(2, -105) > (1, 1001)$ because the first entry, 2, in $(2, -105)$ is larger than the first entry, 1, in $(1, 1001)$. Another example: $(-11, 36) > (-11, -100)$ because, since the first entries in each are equal, both being -11, the second entries then determine the ordering, and 36 is larger than -100.

We leave it as an exercise (see Exercise 6) to the reader to prove:

(i) Given (a, b), (c, d) in R then one and only one of $(a, b) = (c, d)$, $(a, b) \succ (c, d)$, $(c, d) \succ (a, b)$ holds.

(ii) If $(a, b) \succ (c, d)$, $(c, d) \succ (e, f)$, then $(a, b) \succ (e, f)$.

(iii) For any (x, y) ε R whenever $(a, b) \succ (c, d)$ then $(a, b) + (x, y)$ $\succ (c, d) + (x, y)$.

How does this ordering interact with the product in R? Note that $(1, 0) \succ (0, 0)$ and $(0, 1) \succ (0, 0)$, yet $(1, 0) \cdot (0, 1) = (0, 0)$; in other words, the product of two positive elements in R need not be positive.

Exercises

1. In the example above of the ring A consisting of $[0]$, $[1]$, ... , $[5]$, find all values a so that $[a] \cdot [b] = [1]$ for some b.
2. In the ring B consisting of $\{0\}$, $\{1\}$, $\{2\}$, ... , $\{6\}$ given above, prove that if $\{a\} \neq \{0\}$, then $\{a\}^6 = \{1\}$.
3. In Z find all integers m so that $mn = 1$ for some integer n.
4. For the ordering in Z prove, where a, b, c ε Z:
   (a) If $a > b$ and $b > c$, then $a > c$.
   (b) If $a > b$, then $a + c > b + c$ for any c ε Z.
   (c) If $a > b$ and $c > 0$, then $ac > bc$.
   (d) If $c > 0$ and $ac > bc$, then $a > b$.
5. In the lexicographic ordering given on R, is
   (a) $(1, 6) \succ (0, 51)$?
   (b) $(-5, 13) \succ (-4, 13)$?
   (c) $(-73, -62) \succ (-73, -61)$?
6. Prove that in R the statements (i), (ii), and (iii) immediately above are valid.
7. If F is a field and in F, $ab = 0$, prove that at least one of $a = 0$ or $b = 0$ is true.

4. Some Basic Properties

We saw earlier that an attempt to single out the integers by a formal description of their additive and multiplicative properties ended in failure. All we obtained was the collection of commutative rings with unit element. Even conditioning a commutative ring by the presence of an ordering, while it narrowed the field of possibility, was not sufficient to isolate the integers. We now go a step further and introduce an additional restriction that is a fundamental property of the ring of integers.

Assumption L.   Any nonempty set of nonnegative integers has a smallest member.

This is a genuine property that the ring of integers does not share with all commutative rings with an ordering. (In fact, one can show that the addition of assumption L to our previous ones does characterize the integers.) If we consider the example studied earlier, of the ring R of ordered pairs of integers under the lexicographic ordering, then in this example, assumption L does not hold true. In R, let A be the set of all $(1, a)$ with a any integer; every element in A is positive in the ordering of R yet A has no smallest element. If it did it would be of the form $(1, a_o)$; but since $(1, a_o - 1)$ is in A and is smaller than $(1, a_o)$ we get a contradiction of the alleged minimal nature of $(1, a_o)$.

Thus assumption L is not automatically satisfied in rings with an ordering and so it does impose a restriction. We now see how to use it.

Theorem 2.1.   Any integer $a > 1$ is a prime or a product of primes.

Proof.   Suppose the theorem is false. Then there is some integer $b_o > 1$ that is neither a prime nor a product of primes. Let B be the set of all such integers; that is, B is the set of all x with x not a prime nor a product of primes. By assumption, $b_o \; \varepsilon \; B$; hence B is not empty. Thus we have a certain nonempty set B of positive integers; by assumption L, B has a minimal element u. Being in B, u cannot be a prime; thus $u = xy$ where $1 < x < u$, $1 < y < u$. Since x and y are both smaller than u, they *cannot* be in B, hence x is either a prime or a product of primes, as is y. Thus

$$x = p_1 \cdots p_k, y = q_1 \cdots q_m \quad \text{where} \quad p_1, \ldots, p_k, q_1, \ldots, q_m$$

are primes. This yields $u = xy = p_1 \cdots p_k q_1 \cdots q_m$; in other words, u is a product of primes. In that case u cannot be in B. Having obtained both $u \; \varepsilon \; B$ and $u \; \notin \; B$, we have a contradiction and the theorem is proved.

The theorem is usually stated as: Every integer $a > 1$ is a product of primes (understanding a prime to be a product of primes; we shall use it in this sense hereafter). One could ask in how many ways a given integer can be decomposed as such a product. We shall soon see that the answer is: in effect in only one way.

Assumption L gives rise to a very important mathematical principle that is a powerful tool in all of mathematics. This is the *principle of mathematical induction*.

Principle of Mathematical Induction: Let P be a proposition about the positive integers such that

   1. $P(1)$ is true.

2. The truth of $P(m)$ for an integer m implies the truth of $P(m + 1)$. Then P is true for all positive integers.

We now give a proof of this principle on the basis of assumption L.

Let A be the set of positive integers for which P is false. If A is empty, then P is true for all positive integers and we are done. If A is not empty, as a set of positive integers it must have a minimal element m_o. Since $P(1)$ is true, $1 \notin A$; hence $m_o > 1$ and so $m_o - 1$ is positive. Now $m_o - 1$ is not in A since it is smaller than m_o. This means that $P(m_o - 1)$ is true. However, the truth of $P(m_o - 1)$, by assumption 2 on P, implies that of $P((m_o - 1) + 1) = P(m_o)$, hence $m_o \notin A$. Together with $m_o \varepsilon A$, we get a contradiction. Hence no such m_o exists and we get that A is empty.

We give two illustrations of the principle of mathematical induction, the first numerical and the second set-theoretic.

What is the sum of the first n positive integers? We try a few cases: $1, 1 + 2, 1 + 2 + 3, 1 + 2 + 3 + 4$ equal 1, 3, 6, 10, respectively. Now

$$1 = \frac{1(1 + 1)}{2}, \quad 3 = \frac{2(2 + 1)}{2}, \quad 6 = \frac{3(3 + 1)}{2}, \quad 10 = \frac{4(4 + 1)}{2}$$

Let us make a guess that $1 + 2 + \cdots + n = n(n + 1)/2$; how do we prove that this guess is correct? We phrase this in the framework of mathematical induction.

Let $P(n)$ be the proposition which asserts that $1 + 2 + \cdots + n = n(n + 1)/2$. Is $P(1)$ true? In other words, is $1 = 1(1 + 1)/2$? The answer is clearly yes. We now go to the so-called inductive step. Suppose that $P(m)$ is true, that is, that $1 + 2 + \cdots + m = m(m + 1)/2$; does this imply that $P(m + 1)$ is true, that is, that

$$1 + 2 + \cdots + m + (m + 1) = \frac{(m + 1)((m + 1) + 1)}{2}$$
$$= \frac{(m + 1)(m + 2)}{2}$$

Now

$$1 + 2 + \cdots + m + (m + 1) = (1 + 2 + \cdots + m) + (m + 1)$$

By the truth of $P(m)$, $1 + 2 + \cdots + m = m(m + 1)/2$; hence

$$1 + 2 + \cdots + m + m + 1 = \frac{m(m + 1)}{2} + m + 1 = \left(\frac{m}{2} + 1\right)(m + 1)$$
$$= \frac{(m + 1)(m + 2)}{2}$$

Thus the truth of $P(m)$ implies that of $P(m + 1)$. All the conditions of the principle of mathematical induction hold for the proposition P; hence P is true for all positive integers. In other words, this says that $1 + 2 + \cdots + n = n(n + 1)/2$ as claimed.

We now go to our second illustration. In this we redo the enumeration of the number of subsets of a finite set, which we did earlier in Chapter 1, Section 2. Let S be a set having n elements; how many subsets does S have? We claim the answer is 2^n. Now we want to prove it. To do so we phrase it in such a way that it looks like the setup above, in the enunciation of the principle of mathematical induction.

Let $P(n)$ be the proposition that asserts that a set having n elements has 2^n subsets. Is $P(1)$ true? In other words, does a set having 1 element have $2^1 = 2$ subsets? This is clear, for the set $S = \{x_1\}$ has S and ϕ, the empty set, as its only subsets.

We now go to the inductive step. Suppose we know that any set with m elements, for a certain m, has 2^m subsets. Let S have $m + 1$ elements; we write $S = \{x_1, x_2, \ldots, x_m, x_{m+1}\}$. Let $S_o = \{x_1, \ldots, x_m\}$. S_o is a set having m elements; hence S_o has 2^m subsets. Every subset of S_o is a subset of S, so we have now produced 2^m subsets of S. What others exist? Every other subset, as yet not accounted for, must contain the element x_{m+1}. How many such are there? Clearly these are all obtained as follows: Take a subset of S_o and adjoin x_{m+1} to it. This gives rise to 2^m subsets of S, each of which is distinct from the earlier obtained subsets of S_o because it contains x_{m+1}. We have now produced $2^m + 2^m = 2 \cdot 2^m = 2^{m+1}$ subsets of S. These are all, for we have considered all possibilities corresponding to x_{m+1}, lying or not lying in the subset. Thus S has 2^{m+1} subsets. We have shown that the truth of $P(m)$ implies that of $P(m + 1)$. All the conditions for the principle of mathematical induction hold true for the proposition P; hence P is true for all positive integers. Writing what this means we obtain that a set having n elements must have 2^n subsets.

Notice that in both examples given the principle of mathematical induction was a method of proof, not a method of producing an answer. It did not tell us the answer to: What is $1 + 2 + \cdots + n$? Instead, once we guessed a value for $1 + 2 + \cdots + n$ as $n(n + 1)/2$, it allowed us to verify that our guess was correct. So it is fundamentally a method of proof.

Assumption L has another very important implication for the integers, the *Euclidean algorithm*. This algorithm affords us a ready tool to study the divisibility properties of integers. The proof is really by induction. However, what is needed is a slightly sharper version of mathematical induction than the one we just gave. (This form can be found in Exercise 2). So we fall back on direct use of assumption L to carry out the proof of Theorem 2.2 (and Theorems 2.3 and 2.5).

Theorem 2.2. (*Euclidean algorithm*). Given two positive integers a and b, we can find integers q and r such that $a = qb + r$ where $0 \leqq r < b$.

Proof. If $a < b$ the result is correct, for we merely take $q = 0$, $r = a$; then $a = 0 \cdot b + a$ with $0 < a < b$ satisfies the conclusion of the theorem.

Suppose then that $a \geqq b$. Let A be the set of those $a - mb$, where m is an integer, which are nonnegative. Since $a - 1 \cdot b = a - b \geqq 0$, $a - b \, \varepsilon \, A$; hence A is not empty. But then, as a nonempty set of nonnegative integers, A must have a minimal element c. Being in A, c must be of the form $c = a - m_o b$ and $c \geqq 0$. We claim that $c < b$. If not, $c \geqq b$ and hence $c - b \geqq 0$; but $c - b = (a - m_o b) - b = a - (m_o + 1)b$. Thus $0 \leqq a - (m_o + 1)b = c - b < c$. The element $a - (m_o + 1)b$ being nonnegative is in A by the very definition of A; being smaller than c, which was the smallest element in A, it cannot be in A. The only way out of this quandary is that c cannot satisfy $c \geqq b$, hence $c < b$. Now $a = m_o b + (a - m_o b) = m_o b + c$; putting $q = m_o$ and $r = c$, the assertion of the theorem is satisfied.

What this algorithm states is something very familiar to all of us: If you divide one whole number by another you get a remainder which is smaller than the number by which you divide.

Given two integers a and b, by a *common divisor* c of a and b we mean an integer c such that $c|a$ and $c|b$. What should the *greatest common divisor* of two integers be? The "common divisor" part is clear; by the "greatest" part we shall mean a property conveying the idea of "greatest" for the purpose of studying divisibility.

Definition.   The positive integer c is called the *greatest common divisor* of the nonzero integers a and b if

1. $c|a$ and $c|b$
2. whenever $x|a$ and $x|b$, then $x|c$

The mere fact that we have defined what is desired of a greatest common divisor of two integers does not guarantee its existence. Our next result will settle this, but first a word. In the definition we said "*the* greatest common divisor"; this implies a certain uniqueness. We justify this; if c, d were both greatest common divisors of a and b, then $c|d$ and $d|c$ follow, hence $c = \pm d$. But since both c and d are, by hypothesis, positive we do indeed get $c = d$, which is to say, the greatest common divisor (if it exists) is unique. Now to its existence.

Theorem 2.3.   Given two nonzero integers a, b, they have a greatest common divisor c. Moreover, c can be represented as $c = ma + nb$ for an appropriate choice of integers m and n.

Proof.   Let A be the set of all $xa + yb$ with x and y any integers. A contains a and $-a$, one of which is positive. Therefore A contains some positive integers. By assumption L, A has a smallest positive integer c. As an element of A, c has the form $c = ma + nb$.

We claim that $c|t$ for all $t \, \varepsilon \, A$. If $t = xa + yb \, \varepsilon \, A$ is positive, then by the Euclidean algorithm $t = qc + r$ where $0 \leqq r < c$. Writing this explicitly, we get $xa + yb = q(ma + nb) + r$, hence $r = (x - qm)a + (y - qn)b$. From its form, r is in A. Since it is smaller than c, which was the least positive integer in A, we have that r cannot be positive. But since $0 \leqq r$ we conclude that $r = 0$. Thus $t = qc$, whence $c|t$. If $t \, \varepsilon \, A$ is negative, then $-t \, \varepsilon \, A$ and $-t$ is positive so, by what has just been done, $-t$ is a multiple of c, hence t is. If $t = 0$, clearly $t = 0 \cdot c$ so t is a multiple of c.

We assert that c is our required greatest common divisor of a and b. Since $c|t$ for all $t \, \varepsilon \, A$ and $a = 1a + 0b$ and $b = 0a + 1b$ are in A, we deduce that $c|a$ and $c|b$. If x is an integer such that $x|a$ and $x|b$, then $x|(ma + nb)$; hence $x|c$. This completes the proof of the theorem.

We write the greatest common divisor of a and b as (a, b). In case $(a, b) = 1$; that is, a and b have no nontrivial common factor, we call a and b *relatively prime*. Note that as a consequence of the theorem we are able to state the

Corollary. If a and b are relatively prime, then for suitable integers m and n, $ma + nb = 1$.

We leave the following to the reader to verify: An integer $p > 1$ is a prime if and only if for any integer $a \neq 0$ either $p|a$ or $(p, a) = 1$ (Exercise 12).

The above theorem and corollary have a nice implication. Suppose that a and b are relatively prime and that $a|(bc)$; we assert that a must divide c. For, since $(a, b) = 1$, for a suitable choice of m, n we have $ma + nb = 1$, hence $mac + nbc = (ma + nb)c = 1c = c$. Now $a|(mac)$ by definition and $a|(nbc)$ by hypothesis, hence $a|(mac + nbc)$, that is $a|c$. We summarize this as

Theorem 2.4.   If $(a, b) = 1$ and $a|(bc)$, then $a|c$.

It too has two consequences (the second a special case of the first). The proof of the first we leave to the reader (Exercise 13).

Corollary. If p is a prime number and $p|(a_1 a_2 \cdots a_n)$, then p must divide at least one of the a_i.

Corollary. If p is a prime and $p|a^n$, then $p|a$.

We conclude this section by sharpening the statement of Theorem 2.1 from a mere existence statement to one about uniqueness.

Theorem 2.5. Given an integer $a > 1$, a can be factored in one and only one way as

$$a = p_1^{m_1} p_2^{m_2} \cdots p_k^{m_k}$$

where $p_1, \ldots, p_k$ are primes with $p_1 < p_2 < \cdots < p_k$ and the m_i are positive integers.

Proof.   Theorem 2.1 assures us of the possibility of finding such a factorization of a. To establish our present result we must merely prove the uniqueness of the decomposition.

Suppose then that $a > 1$ and

$$a = p_1^{m_1} p_2^{m_2} \cdots p_r^{m_r} = q_1^{n_1} q_2^{n_2} \cdots q_s^{n_s}$$

where $p_1 < p_2 < \cdots < p_r$ and $q_1 < q_2 < \cdots < q_s$ are primes and where $m_1, \ldots, m_r, n_1, \ldots, n_s$ are all positive integers. We want to demonstrate

1. $r = s$
2. $p_1 = q_1, p_2 = q_2, \ldots, p_r = q_r$
3. $m_1 = n_1, m_2 = n_2, \ldots, m_r = n_r$

Let A consist of all integers larger than 1 that have at least two distinct such decompositions. If A is empty, our theorem is proved; if not, A has a least element. We denote this element by a and assume for notational purposes that this is the a we wrote above factored in two ways. If a were a prime it would equal p_1, on the one hand, and q_1 on the other, from the definition of a prime; this would lead us to $a = p_1 = q_1$, contrary to the assumption that the p's and q's gave distinct factorizations of a. Thus a is not a prime and so $b = a/p_1 \neq 1$.

Since $p_1|a$ we must have $p_1|(q_1^{n_1} \cdots q_s^{n_s})$; hence, by the corollary above, $p_1|q_i^{n_i}$ for some i. Therefore $p_1 = q_i \geqq q_1$. Similarly, $q_1 \geqq p_1$. The net result of this is that $p_1 = q_1$. Hence

$$a = p_1^{m_1} p_2^{m_2} \cdots p_r^{m_r} = p_1^{n_1} q_2^{n_2} \cdots q_s^{n_s}$$

Now $b = a/p_1$ is not in A because it is smaller than a. Thus it has a unique factorization of the form given. But

$$b = \frac{a}{p_1} = p_1^{m_1-1} p_2^{m_2} \cdots p_r^{m_r} = p_1^{n_1-1} q_2^{n_2} \cdots q_s^{n_s}$$

The net result of this is that

$$m_1 - 1 = n_1 - 1, m_2 = n_2, \ldots, m_r = n_r, r = s$$

and

$$p_2 = q_2, \ldots, p_r = q_r$$

We already know that $p_1 = q_1$. Finally, from $m_1 - 1 = n_1 - 1$ we obtain $m_1 = n_1$. The unique factorization theorem is now proved.

Exercises

1. Prove the following version of the principle of mathematical induction: Let P be a proposition about the integers such that
   (a) $P(m_o)$ is true
   (b) If $P(m)$ is true for some $m \geq m_o$, then so is $P(m + 1)$.
   Then $P(n)$ is true for all $n \geq m_o$.

2. Prove the following version of the principle of mathematical induction: Let P be a proposition about the integers such that
   (a) $P(1)$ is true.
   (b) If $P(m)$ is true for all positive integers m that are less than s, then $P(m)$ is true.
   Then $P(n)$ is true for all positive integers n.

3. Prove by mathematical induction that
$$1^2 + 2^2 + \cdots + n^2 = \frac{n(n + 1)(2n + 1)}{6}$$

4. Prove by mathematical induction that
$$1^3 + 2^3 + \cdots + n^3 = (1 + 2 + \cdots + n)^2$$

5. We say a set of lines in the plane is *in general position* if no three go through the same point and no two are parallel. Show, using mathematical induction, that n lines in general position cut the plane into $(n^2 + n + 2)/2$ regions.

6. Suppose you have three pegs and on one of these you have piled up n rings of decreasing radius. Using induction prove that to move these n rings from the given peg to another peg, in such a way that no larger ring is ever placed over a smaller one, requires $2^n - 1$ moves. (A move is the shifting of one ring from one peg to another. This is called the problem of the towers of Hanoi.)

7. Using induction prove that the number of distinct ways of filling n slots with n given different objects, one object per slot, is $n!$.

8. Given n objects prove, using induction, that the number of ways of picking pairs of objects from these is $n(n - 1)/2$.

9. Prove that the number of ways of picking groups of i objects from n objects, where $1 \leq i \leq n$, is
$$n(n - 1)(n - 2) \cdots (n - i + 1)/i!$$

   (Another way of saying this is that a set of n elements has $n(n - 1)(n - 2) \cdots (n - i + 1)/i!$ distinct subsets having i elements.)

10. Given two positive integers a and b, the *least common multiple* of a and b, written $[a, b]$, is the positive integer x such that $a|x$ and $b|x$, and x divides every such common multiple. Prove that $[a, b]$ exists and that $[a, b] = ab/(a, b)$.

11. Find the greatest common divisor of 242 and 938.
12. Prove that a positive integer p is a prime if and only if given any integer $a \neq 0$ then (a, p) is either 1 or p.
13. If p is a prime and $p|(a_1a_2 \cdots a_n)$, prove that $p|a_i$ for some i.
14. If $(a, b) = 1$ (a and b are relative prime) and a and b both divide c, prove that $(ab)|c$.
15. If a and b are factored into primes as $a = p_1^{\alpha_1} \cdots p_k^{\alpha_k}$, $b = p_1^{\beta_1} \cdots p_k^{\beta_k}$ where $p_1, \ldots, p_k$ are primes and $\alpha_1, \ldots, \alpha_k, \beta_1, \ldots, \beta_k$ are nonnegative integers, express (a, b) in terms of the p's, α's, and β's.
16. Show that the following is a method for computing (a, b) where a, b are positive integers: Write

$$a = q_1b + r_1, \quad 0 \leq r_1 < b$$
$$b = q_2r_1 + r_2, \quad 0 \leq r_2 < r_1$$
$$r_1 = q_3r_2 + r_3, \quad 0 \leq r_3 < r_2$$
$$\cdot \quad \cdot \quad \cdot$$

Since $r_1 > r_2 > r_3 > \cdots \geq 0$, prove that you must reach 0 at some first r_k. Then show that $r_{k-1} = (a, b)$. (This is the form in which the algorithm appears in Euclid's famous *Elements*.) (Illustration: $a = 219$, $b = 15$.

$$219 = 14 \cdot 15 + 9, \quad r_1 = 9$$
$$15 = 1 \cdot 9 + 6, \quad r_2 = 6$$
$$9 = 1 \cdot 6 + 3, \quad r_3 = 3$$
$$6 = 2 \cdot 3 + 0, \quad r_4 = 0$$

so $r_3 = 3$ is $(219, 15)$.)

In Exercises 17 through 21 S will be the set of all integers larger than 1 of the form $4n + 1$. Thus $S = \{5, 9, 13, 17, 21, \ldots\}$. Call an integer $m \, \varepsilon \, S$ a "prime object" of S if we *cannot* factor m as $m = ab$ with both a and b in S.

17. (a) Show that 9, 21, 39, 49 are "prime objects" of S.
    (b) Show that we can factor 441, which is in S, as the product of "prime objects" of S, in two distinct ways.
18. (a) Show that every prime number of the form $4n + 1$ is a "prime object" of S.
    (b) If p and q are any two prime numbers of the form $4n + 3$, show that pq is a "prime object" of S. (Note: We do not exclude the possibility that $p = q$. See Exercise 17(a).)
    (c) Show that every "prime object" of S is of the form described in (a) or (b).
19. Show that every element in S can be factored as a product of "prime objects" of S.
20. Let $m \, \varepsilon \, S$; as an integer, we can factor m as $m = p_1^{\alpha_1} \cdots p_k^{\alpha_k}$

$q_1 q_2 \cdots q_t$, where $p_1, \ldots, p_k$ are distinct primes of the form $4n + 1$ and $q_1, \ldots, q_t$ are primes (not necessarily distinct) of the form $4n + 3$. Show that we can factor m in a unique way as a product of powers of "prime objects" of S if and only if $t < 4$. (Thus, for instance, $825 = 5^2 \cdot 3 \cdot 11$, for which t is 2, can be factored in a unique way as a product of powers of "prime objects" of S. On the other hand, $441 = 3 \cdot 3 \cdot 7 \cdot 7$, for which $t = 4$, can be factored in more than one way as a product of powers of "prime objects" of S.)

21. If a and b are in S, we say that c in S is a greatest common divisor of a and b (relative to S) if:
    (a) $c|a$ and $c|b$
    (b) Whenever $u \ \varepsilon \ S$ divides both a and b, then u divides c. Do any two integers in S have a greatest common divisor relative to S?

22. This exercise outlines an alternate proof of uniqueness of prime factorization in the integers. Suppose that uniqueness fails, and let n be the smallest integer with two distinct factorizations: $n = p_1 \cdots p_r = q_1 \cdots q_s$. Argue that the p's must all be distinct from the q's. Pick p_1 minimal among the p's, q_1 minimal among the q's. We have that $n - p_1 q_1$ is positive and less than n. In the unique factorization of $n - p_1 q_1$, both p_1 and q_1 must appear. Thus $n - p_1 q_1$ is divisible by $p_1 q_1$, and so is n, a contradiction to the uniqueness of factorization of n/p_1.

5. Equivalence Relations

We should like to broaden the notion of equality for integers to be able to discuss equality "up to a point." Because the upcoming discussion is valid in a wider setting than that of the integers alone, we briefly digress to the framework of sets in general.

Let S be any set. We have an intuitive idea of what it means to say that there is a relation defined among the elements of S. This concept could be formalized, and defended vigorously, but our vague, intuitive feeling about it will suffice in what we are about to do.

A relation, $\sim$, on S is called an *equivalence relation* on S if:

1. $a \sim a$ (reflexivity)
2. $a \sim b$ implies $b \sim a$ (symmetry)
3. $a \sim b$, $b \sim c$ imply $a \sim c$ (transitivity)
   for any $a, b, c \ \varepsilon \ S$.

Clearly this notion generalizes that of equality, for the relation "a is identical to b" is an equivalence relation. Before proceeding with the discussion of such relations, we look at some examples. In each of the following, the relation ∼ defined on the particular set S is an equivalence relation.

1. Let S be the set of all people, and let $a \sim b$ mean that a and b have the same father.
2. Let S be the set of all objects for sale in a given department store; for $a, b \; \varepsilon \; S$ let $a \sim b$ if a and b sell for the same price.
3. Let S be the set of integers; for $a, b \; \varepsilon \; S$ let $a \sim b$ if $a^2 = b^2$.
4. Let S be any set and let f be a mapping from S to some set T. We define, for $a, b \; \varepsilon \; S$, that $a \sim b$ if $f(a) = f(b)$.
5. Let S be the set of integers; for $a, b \; \varepsilon \; S$ define $a \sim b$ if $a - b$ is even.
6. Let S be the set of integers and let $n > 1$ be a fixed integer. For $a, b \; \varepsilon \; S$ define $a \sim b$ if $n|(a - b)$.

Not all relations on a set are equivalence relations. For instance, the following are not.

(i) Let S be the set of all people. The relation: "a is the brother of b" is not an equivalence relation. (Show that it fails to satisfy all three properties.)
(ii) Let S be the set of integers. The relation: "a is relatively prime to b" is not an equivalence relation.

We make a few remarks about the examples given above. First and foremost is the observation that in all of the examples we fix on some attribute of the elements of some set and measure equality up to this attribute, ignoring all other behavior of the elements. This is what was meant earlier by saying "equality up to a point." In discussing equality up to a point we blot out some differences that may not be relevant to a particular discussion in hand.

Thus in Example 1 we lump people together according to the criterion that they have the same father, ignoring, for instance, whether they are male or female, old or young, or whether they have the same mother. In Example 2, we ignore the quality, utility, etc., of the objects in the department store; for us, under the criterion of equivalence set forth, all that counts is the price of the object. In Example 3, by defining two integers as equivalent if they have the same square, we ignore the sign of an integer, contenting ourselves only with its size. Example 4 is very interesting, for, as we shall see later, it is the universal description of any equivalence relation; by this we mean the following: Given a set S with an equivalence relation, we can find some set T and some mapping f from S to T such that $a \sim b$ if and only if $f(a) = f(b)$. This remark is not very deep, for we shall soon see how to construct the set T and the mapping f. In Example 5, we are fixing on the parity of an integer, using it as the criterion for "equal up to a point." Clearly Example 5 is a special case of Example 6 with

$n = 2$. In fact, it is because of Example 6 that we have decided to discuss equivalence relations here. We shall go into a more detailed study of Example 6 in the next section.

Now that we have the notion of an equivalence relation, what can we do with it? We first introduce a derived concept, that of *equivalence class*.

Definition. If $\sim$ defines an equivalence relation on S, then for $a \ \varepsilon \ S$ the *equivalence class* of a is defined as the set of all x in S with $a \sim x$.

We shall write the equivalence class of a as $[a]$. Note that $[a]$ is a *subset* of S.

Lemma 2.3. Given $a, b \ \varepsilon \ S$, S a set with equivalence relation $\sim$, then either $[a] = [b]$ or $[a]$ and $[b]$ have no element in common.

Proof. To establish the result, it is enough to show that if $[a]$ and $[b]$ do have an element in common then $[a] = [b]$. Because $[a]$, $[b]$ are sets, in order to prove $[a] = [b]$ we merely establish that

$$[a] \subset [b] \quad \text{and} \quad [b] \subset [a]$$

Suppose that $x \ \varepsilon \ [a] \cap [b]$. Since $x \ \varepsilon \ [a]$ we have $a \sim x$; since $x \ \varepsilon \ [b]$ we have $b \sim x$ and so $x \sim b$ by symmetry. From $a \sim x$, $x \sim b$ we deduce $a \sim b$ by transitivity. Let $y \ \varepsilon \ [b]$; then $b \sim y$. However, since $a \sim b$ and $b \sim y$ we get $a \sim y$, which is to say $y \ \varepsilon \ [a]$. We have shown that every $y \ \varepsilon \ [b]$ is in $[a]$; that is, $[b] \subset [a]$. Since the reasoning is clearly symmetric, we similarly get $[a] \subset [b]$. These two opposite containing relations give us the desired outcome $[a] = [b]$.

The result of Lemma 2.3 enables us to elaborate and clinch a remark made earlier. We had said that given any set S with an equivalence relation $\sim$ we can find a set T and a mapping $f: S \to T$ such that $a \sim b$ if and only if $f(a) = f(b)$. We do this by picking for T the set of all $[a]$, and defining the mapping $f: S \to T$ by $f(a) = [a]$. Note that T is a set whose *elements* are subsets of S; f merely assigns to every element in S its equivalence class. We leave it to the reader to verify that this f and T have the requisite properties.

The equivalence classes of the equivalence relation on S furnish us with a nice decomposition of S. As we saw in Lemma 2.3, the equivalence classes are *mutually disjoint* in the sense that two that are not equal have no element in common. Given any a in S, $a \ \varepsilon \ [a]$. Thus the union of all the equivalence classes gives us all of S. We have proved a substantial part (it is merely a paraphrase of Lemma 2.3) of

Theorem 2.6. Given a set S with an equivalence relation $\sim$ then $\sim$ induces a decomposition of S into a union of mutually disjoint nonempty subsets. Conversely, such a decomposition of S enables us to define an equivalence relation on S whose equivalence classes are precisely the pieces in the given decomposition.

Proof.   We have already proved the first part, when we observed that the equivalence classes give us the desired decomposition of S.

Suppose, on the other hand, that S is the union of mutually disjoint nonempty subsets S_α. We declare two elements a, b ε S to be equivalent if they are in the same S_α. Since the S_α are mutually disjoint, it is easy to verify that the relation so defined is an equivalence relation on S and that each S_α is the equivalence class of any element in it.

This theorem is very interesting in the special case when S is a finite set, for there it acquires a "bookkeeping" flavor. We explain. Let S have n elements and suppose that $\sim$ is an equivalence relation on S with equivalence classes $S_1, \ldots, S_k$, having $n_1, n_2, \ldots, n_k$ elements, respectively; then, by the theorem, $n = n_1 + n_2 + \cdots + n_k$. If we had a method of calculating the n_i (clearly this would depend completely on the specific nature of $\sim$) we would get a nice identity. This method of counting up in two different ways is a powerful tool in finite mathematics. We illustrate its use in an example below. Later (e.g., in Lagrange's theorem) we shall see more genuine and striking applications of these ideas.

Let A be a set having m elements and let S be the collection of all subsets of A. As we saw earlier, S has 2^m elements. In S we define an equivalence relation $\sim$ by: If X, Y ε S (remember X, Y are subsets of A), then $X \sim Y$ if and only if X and Y have the same number of elements. Let S_o denote the equivalence class of an element in S having 0 elements, S_1 that of an element in S having 1 element, $\ldots$, S_i the equivalence class of an element in S having i elements.

How many elements are there in S_i; that is, how many subsets are there of A having i elements? For S_o the answer is clear; only the empty subset ϕ of A has 0 elements so S_o has 1 element. For S_1 the question is almost equally easy; the number of subsets of A that have exactly 1 element is m. In fact as a result of Exercise 9 in Section 4, S_i has $m(m-1)\cdots(m-i+1)/i!$ elements. Using Theorem 2.6 (that is, invoking our bookkeeping principle) we get

$$2^m = 1 + m + \frac{m(m-1)}{2} + \frac{m(m-1)(m-2)}{3!}$$
$$+ \cdots + \frac{m(m-1)(m-2)\cdots(m-i+1)}{i!} + \cdots$$

It is always nice to see a general result checked out in a specific case. Take $m = 5$; we should have

$$2^5 = 1 + 5 + \frac{5 \cdot 4}{2} + \frac{5 \cdot 4 \cdot 3}{1 \cdot 2 \cdot 3} + \frac{5 \cdot 4 \cdot 3 \cdot 2}{1 \cdot 2 \cdot 3 \cdot 4} + \frac{5 \cdot 4 \cdot 3 \cdot 2 \cdot 1}{1 \cdot 2 \cdot 3 \cdot 4 \cdot 5}$$

Doing the arithmetic we see that it does indeed check out: $32 = 1 + 5 + 10 + 10 + 5 + 1$. The result above is the special case of the binomial theorem where one expands $(1 + 1)^m$.

Exercises

1. Let S be the set of rational numbers; for $a, b \, \varepsilon \, S$ define $a \sim b$ if $a - b$ is an integer. Prove that this is an equivalence relation on S. What is the equivalence class of 16? What is the equivalence class of $-\frac{3}{2}$?

2. Let S be the plane and P_o a given point in S. Declare $a \sim b$, for $a, b \, \varepsilon \, S$, if the distance from a to P_o equals that from b to P_o. Prove that this is an equivalence relation on S and that the equivalence classes are the circles in the plane having P_o as center.

3. Let S be a nonempty set and suppose that on S there is defined a relation Ω such that for all $a, b, c \, \varepsilon \, S$:
   (a) Given a there exists an $a' \, \varepsilon \, S$ so that $a \, \Omega \, a'$.
   (b) $a \, \Omega \, b$ implies $b \, \Omega \, a$.
   (c) $a \, \Omega \, b$, $b \, \Omega \, c$ implies $a \, \Omega \, c$.
   Prove that Ω is an equivalence relation on S. (Note: Show that the properties given for Ω force $a \, \Omega \, a$ for every $a \, \varepsilon \, S$.)

4. Let S be the set of nonzero rational numbers. If $a, b \, \varepsilon \, S$, define $a \sim b$ if $ab > 0$. Prove that this is an equivalence relation on S. How many equivalence classes are there?

5. If S is the set of integers and $\sim$ is defined on S by $a \sim b$ if $a - b$ is even, prove that this is an equivalence relation on S and determine the equivalence classes.

6. Let S be the set of nonzero rational numbers. If $a, b \, \varepsilon \, S$, declare $a \sim b$ if a/b is an integer. Prove that this is not an equivalence relation on S.

7. Let S be the set of nonzero real numbers. If $a, b \, \varepsilon \, S$, declare $a \sim b$ if a/b is a rational number. Prove that this is an equivalence relation on S.

8. Let S be the set of integers. Where does the relation $a \geq b$ fail to satisfy the requirements for an equivalence relation?

6. Congruence

The discussion in the preceding section took place in the framework of a general set. We particularize it now to a specific relation in the set of integers. In fact the relation to be examined was already introduced in Example 6, illustrating the concept of an equivalence relation.

Let $n > 1$ be an integer that we shall consider as fixed in the discussion to follow. For two integers a and b we define: $a \equiv b$ mod n if $n|(a - b)$. We read this as: "a is congruent to b modulo n" and call the relation "congruence modulo n." We assert that this is indeed an equivalence relation on the set of integers. To verify this assertion we check the three properties of an equivalence relation in turn.

1. $a \equiv a$ mod n for, $n|(a - a) = 0$.
2. $a \equiv b$ mod n implies $b \equiv a$ mod n for, if $n|(a - b)$, then $n|(b - a)$ $= -(a - b)$.
3. If $a \equiv b$ mod n and $b \equiv c$ mod n, then $n|(a - b)$ and $n|(b - c)$; hence $n|\{(a - b) + (b - c)\} = a - c$. Therefore $a \equiv c$ mod n follows.

We have now shown that congruence modulo n is an equivalence relation on the set of integers. In keeping with our notation of the preceding section we write the equivalence class of a as $[a]$. Note that $[a]$ is the set of all integers $a + kn$, $k = 0, \pm1, \pm2, \ldots$ Given any integer a, by Euclid's algorithm we can write a as $a = qn + r$ where q, r are integers and $0 \leq r < n$. Since $n|(a - r)$, $a \equiv r$ mod n; this tells us that $[a] = [r]$. Thus when we list the congruence classes $[0], [1], \ldots, [n - 1]$ we account for *every possible* congruence class. Hence there are at most n distinct congruence classes modulo n. Moreover, we claim that these enumerated classes, namely $[0], [1], \ldots, [n - 1]$, are all different. If not, two of them would be equal. But what would this say? It would say that we could find integers i and j satisfying $0 \leq i < j \leq n - 1$ and $[i] = [j]$. However, we are now placed in the following impossible position: From $0 \leq i < j \leq n - 1$ we deduce that $0 < j - i < n$ and from $[i] = [j]$ that $n|(j - i)$. This is impossible for n cannot divide a positive integer smaller than n. The only way out is the desired conclusion that the n congruence classes $[0], [1], \ldots, [n - 1]$ are all distinct. We summarize what we have shown so far in the discussion as

Theorem 2.7. The relation $a \equiv b$ mod n is an equivalence relation on the set of integers. It has the n distinct equivalence classes $[0], [1], \ldots, [n - 1]$.

The integers are endowed with a multiplication and an addition. How does this new relation we have just introduced intermesh with these operations on the integers? The answer is provided in

Lemma 2.4. If $a \equiv b$ mod n and $c \equiv d$ mod n, then

1. $a + c \equiv b + d$ mod n
2. $ac \equiv bd$ mod n

Proof.   The proof of part 1 is the easier of the two so we prove only part 2 leaving the details of verifying the other to the reader (Exercise 1).

Suppose that $a \equiv b$ mod n and $c \equiv d$ mod n; then $n|(a - b)$ and $n|(c - d)$. Now $ac - bd = (a - b)c + (c - d)b$. Because n divides $a - b$, it divides $(a - b)c$; similarly $n|(c - d)b$. Therefore $n|\{(a - b)c + (c - d)b\} = ac - bd$. By definition this implies that $ac \equiv bd$ mod n.

This lemma asserts that the relation "congruence mod n" respects the addition and multiplication of integers. It tells us that congruence is closer to equality than just any equivalence relation, for congruence obeys the analogs of "equals added to equals are equal" and "equals multiplied by equals are equal." How far can this analogy be pushed? In studying the equality of integers we are also able to cancel a nonzero factor; that is, if $ab = ac$ and $a \neq 0$, then $b = c$. Can we say the same thing for congruence mod n? The answer is no. Note that $5 \cdot 4 \equiv 2 \cdot 4$ mod 12, and yet $5 \not\equiv 2$ mod 12 and $4 \not\equiv 0$ mod 12. It is reasonable to investigate when such cancellation is possible. The purpose of the next lemma is to give us such a criterion.

Lemma 2.5.   If $ab \equiv ac$ mod n and a is relatively prime to n, then $b \equiv c$ mod n.

Proof.   Let us recall that the statement a is relatively prime to n means that their greatest common divisor, (a, n) is 1. From $ab \equiv ac$ mod n we have $n|(ab - ac)$, and so $n|a(b - c)$. Since $(a, n) = 1$, by invoking Theorem 2.4 we get $n|(b - c)$. Hence $b \equiv c$ mod n.

We turn for a short while in another direction. Admittedly many mathematical ideas and techniques are difficult, even for a professional mathematician. However, every once in a while one runs across a mathematical principle which is trivial to enunciate but which nevertheless gives rise to a series of nontrivial results. We introduce one such idea now. It states nothing more profound than the following: If a mailman delivers 51 letters to an apartment house having 50 apartments then some apartment receives at least 2 letters. We formalize this as the *Pigeon-Hole Principle*: If n objects are distributed over m positions, and $n > m$, then some position receives at least two objects.

An equivalent formulation is the following: If n objects are distributed over n positions in such a way that no position gets more than one object, then each position gets exactly one object.

We shall give a number of applications and illustrations of this principle later. We exploit it immediately in studying the integers mod n.

Let a be an integer relatively prime to n. We consider the remainders of a, a^2, a^3, ..., a^{n+1} on division by n. We have written $n + 1$ numbers; their remainders on division by n must fall in the range from 0 to $n - 1$ inclusive. Hence there are n possible remainders for these $n + 1$ integers. By the Pigeon-Hole Principle, two of these numbers must have the same remainder on division by n. But then their difference must be divisible by n. More formally, there must be two such numbers, that is, numbers of the form a^i, a^j with $1 \leq i, j \leq n + 1$ that satisfy $a^i \equiv a^j \bmod n$ although $i \neq j$. Since $i \neq j$, one of these is smaller than the other and we may assume that $i < j$. Because a is relatively prime to n we can use Lemma 2.5 to cancel off a's. The upshot of this is that $a^{j-i} \equiv 1 \bmod n$. What do we know about $j - i$? Since $i < j$, $j - i$ is positive. Since j is at most $n + 1$ and i is at least 1, we have that $j - i \leq n$. Let $k = j - i$; the net result of the discussion is that $a^k \equiv 1 \bmod n$. We state this result as

Lemma 2.6.   If a is relatively prime to n, there exists a positive integer k that satisfies $k \leq n$ and $a^k \equiv 1 \bmod n$.

We shall try to find more precise information about k. This will lead us to some classical theorems in number theory. Note that Lemma 2.6 implies that if a is relatively prime to n then we can find something like a multiplicative inverse for a modulo n; more exactly, if $(a, n) = 1$, there is an integer b ($b = a^{k-1}$ will do) such that $ab \equiv 1 \bmod n$. (Note: Another way to find such an element b is to use the corollary to Theorem 2.3.) We leave it as an informal exercise to prove that if $(a, n) \neq 1$, then it is impossible to find an integer b such that $ab \equiv 1 \bmod n$.

Let us look at an example. We take $n = 10$ and $a = 3$; then we surely have that a is relatively prime to n. When we look at $3^4 = 81$ we see that $3^4 \equiv 1 \bmod 10$. Here 4 plays the role of the k of the lemma.

We now translate everything we have just spoken about into the language of congruence classes. Let Z_n denote the set of congruence classes modulo n; that is $Z_n = \{[0], [1], \ldots, [n - 1]\}$. In Z_n we introduce an addition and a multiplication. (On pp. 53–54 we discussed cases $n = 6$ and $n = 7$; now we shall handle the general case.) We define:

$$[a] + [b] = [a + b]$$
$$[a][b] = [ab]$$

We claim that these operations are meaningful (i.e., well-defined); that is, if $[a] = [a_1]$ and $[b] = [b_1]$, then $[a] + [b] = [a_1] + [b_1]$ and $[a][b] = [a_1][b_1]$. What is involved here? All that is needed is the following simple verification, using Lemma 2.4. Since $[a] + [b] = [a + b]$ and $[a_1] + [b_1] = [a_1 + b_1]$ we must show that $[a + b] = [a_1 + b_1]$. Now since $[a] = [a_1]$, $a \equiv a_1 \bmod n$; similarly $b \equiv b_1 \bmod n$. Hence by Lemma 2.4, $a + b \equiv a_1 + b_1 \bmod n$ whence $[a + b] = [a_1 + b_1]$. Similarly we can show that $[a][b] = [a_1][b_1]$. Therefore the operations are well-defined in

Z_n. We claim that under these operations Z_n is a commutative ring with unit, as defined in Section 2. This can really be seen quite easily.

Let one of the axioms (1) through (5), (i) through (iv), and (D) be presented for verification in Z_n; for example, let us consider (D). Now we do know that the law in question holds in Z; that is,

$a(b + c) = ab + ac$

is valid for any *integers a, b,* and *c*. What we have to see is that the same thing is true when each of the letters is draped in brackets:

$[a]([b] + [c]) = [a][b] + [a][c]$

By several uses of the very definitions of addition and multiplication in Z_n, the second of these equations follows directly from the first. In other words, (D) in Z_n is deducible in an automatic way from (D) in Z. In the same way, every law in Z_n merely mirrors the corresponding law in Z.

An additional word of explanation might be helpful. When we introduced Z_6 and Z_7 in Section 3, we treated the symbol $[a]$ in a somewhat different way. There it actually meant an integer in the range 0 to $n - 1$; for instance, with $n = 6$, $[a]$ took on the values 0, 1, . . . , 5. In our present context, $[a]$ is a *set* of integers, namely the congruence class of a modulo n. We now see that this switch in strategy is a good idea; for instance, it simplifies the task of checking that Z_n is a commutative ring with unit.

Z_n is thus a commutative ring with unit. It is a ring with n elements and is called the *ring of integers mod n*. Speaking about Z_n and about congruence mod n amounts to the same thing.

Let us use the shorthand $[a]^2$ for $[a][a]$, $[a]^3$ for $[a][a][a]$, and in general $[a]^t$ for the result of multiplying together t terms that are all equal to $[a]$. Note that $[a]^t = [a^t]$. Lemma 2.6 then becomes: If $(a, n) = 1$, then $[a]^k = [1]$ for some k satisfying $0 < k \leqq n$; also, if $(a, n) = 1$, there is a b such that $[a][b] = [1]$. We call $[b]$ the *inverse* of $[a]$.

Let p be a prime number and let us consider Z_p, the integers mod p. To say that $(a, p) = 1$ is the same as saying $p \nmid a$; that is, $[a] \neq [0]$. Hence Lemma 2.6 becomes in Z_p, p a prime, if $[a] \neq [0]$, then $[a]^k = [1]$ for some k with $0 < k \leqq p$. Furthermore, a nonzero element in Z_p has an inverse under multiplication in Z_p.

We now change direction a little and make a computation in a particular Z_p, namely when $p = 7$. The nonzero elements in Z_7 are $[1], [2], \ldots,$ $[6]$. We calculate

$[3][1] = [3]$
$[3][2] = [6]$
$[3][3] = [9] = [2]$
$[3][4] = [12] = [5]$
$[3][5] = [15] = [1]$
$[3][6] = [18] = [4]$

Therefore

$$([3][1])([3][2])([3][3])([3][4])([3][5])([3][6]) = [3][6][2][5][1][4]$$

Because multiplication in Z_7 is associative and commutative, we get from this: $[3]^6[1][2] \cdots [6] = [1][2] \cdots [6]$. But this implies that $[3]^6 = [1]$ or, equivalently, that $3^6 \equiv 1$ mod 7.

We imitate this argument in Z_p where p is any prime. Let us see what we get. Let $[1], [2], \ldots, [p-1]$ be the nonzero elements in Z_p and suppose that $[a] \neq [0]$. Now $[a][1], [a][2], \ldots, [a][p-1)$ are all in Z_p and are not $[0]$. Moreover, we claim that these are all distinct. Why? If $[a][i] = [a][j]$, then $[i] = [j]$, by Lemma 2.5. Hence $[a][1], \ldots, [a][p-1]$ gives us $p-1$ distinct nonzero elements in Z_p. *But Z_p has in all only $p-1$ distinct nonzero elements.* By the Pigeon-Hole Principle we must get *all* the nonzero elements in Z_p. Therefore $[a][1], \ldots, [a][p-1]$ are merely the elements $[1], \ldots, [p-1]$ in some order. Since multiplication in Z_p is commutative and associative, taking products we get

$$([a][1])([a][2]) \cdots ([a][p-1]) = [1][2] \cdots [p-1]$$

and so

$$[a]^{p-1}[1][2] \cdots [p-1] = [1][2] \cdots [p-1]$$

Since $1, 2, \ldots, p-1$ are relatively prime to p, using Lemma 2.5 we can cancel these; this leaves us with $[a]^{p-1} = [1]$, which is to say $[a^{p-1}] = [1]$. In terms of congruences this translates into $a^{p-1} \equiv 1$ mod p. We have proved a very famous old result due to Fermat known as the *little Fermat theorem*:

Theorem 2.8. If p is a prime and a is an integer not divisible by p, then $a^{p-1} \equiv 1$ mod p.

> P. Fermat (1601–1665) is one of the great number theorists of all time. This famous French mathematician was an "amateur" in that he practiced law for a living; he was an amateur only in that sense. He is also one of the founders of probability theory. The famous Fermat last problem, namely that $x^m + y^m = z^m$ has no solution in positive integers for $m > 2$, is still an open question. Attempts to solve Fermat's last problem gave rise to the creation of algebraic number theory and much of modern algebra. A forthcoming book on Fermat by Michael Mahoney is previewed in an article in the October 6, 1972, issue of *Science*.

Try this out with $p = 11$ and $a = 3$. (Hint: Don't compute 3^{10} in full; compute early powers mod 11.)

In the course of the proof of the little Fermat theorem the product $[1][2] \ldots [p-1]$ arose; we had no need of its value, so we made no attempt to evaluate it. Now we turn to a calculation of this number. What this really amounts to is asking: To what is $(p-1)!$ congruent for p a prime?

Let us experiment with the early primes.

$p = 3, (3 - 1)! = 1 \cdot 2 \equiv -1 \bmod 3$

$p = 5, (5 - 1)! = 4! = 24 \equiv -1 \bmod 5$

$p = 7, (7 - 1)! = 6! = 720 \equiv -1 \bmod 7$

What would be a reasonable guess for the general result? Clearly, $(p - 1)!$ $\equiv -1 \bmod p$ when p is a prime. Before going on to establish the result, we do it for $p = 11$, not by a brute force calculation of 10!, but rather by a more subtle playing around.

For $p = 11$ we know that every nonzero element in Z_{11} has a multiplicative inverse in Z_{11}. We write these inverses down.

$1 \cdot 1 \equiv 1 \bmod 11$

$2 \cdot 6 = 6 \cdot 2 \equiv 1 \bmod 11$

$3 \cdot 4 = 4 \cdot 3 \equiv 1 \bmod 11$

$5 \cdot 9 = 9 \cdot 5 \equiv 1 \bmod 11$

$7 \cdot 8 = 8 \cdot 7 \equiv 1 \bmod 11$

$10 \cdot 10 = 100 \equiv 1 \bmod 11$

Notice that outside of 1 and 10, every element is different from its inverse.

Now $10! = 1 \cdot 2 \cdot 3 \cdot 4 \cdot 5 \cdot 6 \cdot 7 \cdot 8 \cdot 9 \cdot 10$; we write this as $10!$ $= 1 \cdot (2 \cdot 6) \cdot (3 \cdot 4) \cdot (5 \cdot 9) \cdot (7 \cdot 8) \cdot 10$. Going modulo 11, and using that $2 \cdot 6 \equiv 1 \bmod 11$, $3 \cdot 4 \equiv 1 \bmod 11$ and so on, we get that $10! \equiv 10 \equiv -1 \bmod 11$.

We shall pattern the argument for the general result after that given above for 11. The result is a famous old theorem called *Wilson's theorem*.

Theorem 2.9. If p is a prime number, then $p - 1! \equiv -1 \bmod p$.

J. Wilson (1741–1793) was an English mathematician remembered only for the theorem proved here that bears his name.

Proof. Let p be a prime number; we consider $(p - 1)! = 1 \cdot 2 \cdot 3 \cdots$ $(p - 1)$. As a consequence of Lemma 2.6, for any integer a in the range from 1 to $p - 1$ we can find $f(a)$ in the same range such that $af(a) \equiv 1 \bmod p$. Call $f(a)$ the *relative inverse* of a. For what values of a is $f(a)$ equal to a? Equivalently, when is $a^2 \equiv 1 \bmod p$?

If $a^2 \equiv 1 \bmod p$, then $p|(a^2 - 1)$; however $a^2 - 1 = (a - 1)$ $(a + 1)$, hence since $p|(a - 1)(a + 1)$ and p is a prime, by the corollary to Theorem 2.4 we must have that $p|(a - 1)$ or $p|(a + 1)$. These latter relations give $a \equiv 1 \bmod p$ or $a \equiv -1 \bmod p$; in short, if $a^2 \equiv 1 \bmod p$, then $a \equiv 1 \bmod p$ or $a \equiv -1 \bmod p$.

Thus in the expression $1 \cdot 2 \cdot 3 \cdots (p - 1) = (p - 1)!$ the terms at the ends are the only ones equal to their own relative inverses. Therefore for $a = 2, 3, \ldots, p - 2$, $f(a) \neq a$. We write out $2 \cdot 3 \cdots \cdots (p - 2)$ by pairing each a with its $f(a)$ (recall what we did for $p = 11$). Going modulo p and using $af(a) \equiv 1 \bmod p$, we get

$2 \cdot 3 \cdots (p-2) \equiv 1 \bmod p$. Consequently, $(p-1)! = 1 \cdot 2 \cdot 3 \cdot \cdots (p-2)(p-1) \equiv -1 \bmod p$. This is precisely the result claimed in the theorem. Wilson's theorem is now proved.

From Wilson's theorem we are able to deduce some information about square roots modulo a prime. First we consider two examples.

Let $p = 11$ and consider $((p-1)/2)! = ((11-1)/2)! = 5!$. Working modulo 11 we see that $5! = 120 \equiv -1 \bmod 11$, and so $(5!)^2 \equiv 1 \bmod 11$. Now consider $p = 13$; we calculate $((p-1)/2)! = ((13-1)/2)! = 6! = 720$. Going modulo 13 we get $6! \equiv 5 \bmod 13$ and then $(6!)^2 \equiv -1 \bmod 13$.

Note the difference in these two cases. In the first, going to the halfway point of p (namely to $(p-1)/2$), taking factorial, squaring, and going modulo p, we got $+1$ as the answer; in the second case we got -1 as the answer. Why the difference in sign?

Instead of immediately proceeding to the general case we examine these two special cases a little more closely. According to Wilson's theorem $1 \cdot 2 \cdot 3 \cdot 4 \cdot 5 \cdot 6 \cdot 7 \cdot 8 \cdot 9 \cdot 10 = 10! \equiv -1 \bmod 11$. We split this in the middle as $(1 \cdot 2 \cdot 3 \cdot 4 \cdot 5)(6 \cdot 7 \cdot 8 \cdot 9 \cdot 10)$. Now $10 \equiv -1 \bmod 11$, $9 \equiv -2 \bmod 11$, $8 \equiv -3 \bmod 11$, $7 \equiv -4 \bmod 11$, and $6 \equiv -5 \bmod 11$; therefore $6 \cdot 7 \cdot 8 \cdot 9 \cdot 10 \equiv (-5)(-4)(-3)(-2)(-1) = -5! \bmod 11$. Note that the minus sign came about because in the product $6 \cdot 7 \cdot 8 \cdot 9 \cdot 10$ there is an *odd* number of terms. Note further that this odd number is merely the number of integers from 1 to 5; that is $(11-1)/2$. Going back to

$$-1 \equiv 10! = (1 \cdot 2 \cdot 3 \cdot 4 \cdot 5)(6 \cdot 7 \cdot 8 \cdot 9 \cdot 10)$$
$$\equiv (1 \cdot 2 \cdot 3 \cdot 4 \cdot 5)(-(1 \cdot 2 \cdot 3 \cdot 4 \cdot 5)) \equiv -(5!)^2 \bmod 11$$

we see why $(5!)^2 \equiv 1 \bmod 11$.

We now do a similar thing for $p = 13$. From $12! \equiv -1 \bmod 13$ we get $(1 \cdot 2 \cdot 3 \cdot 4 \cdot 5 \cdot 6)(7 \cdot 8 \cdot 9 \cdot 10 \cdot 11 \cdot 12) \equiv -1 \bmod 13$. But $12 \equiv -1 \bmod 13$, $11 \equiv -2 \bmod 13$, etc. gives us $7 \cdot 8 \cdot 9 \cdot 10 \cdot 11 \cdot 12 \equiv (-6)(-5)(-4)(-3)(-2)(-1) = +6! \bmod 13$. Here we get $+6!$ because the number of minuses, namely $6 = (13-1)/2$, is even. Thus $-1 \equiv 12! \equiv (1 \cdot 2 \cdot 3 \cdot 4 \cdot 5 \cdot 6)^2 = (6!)^2 \bmod 13$.

Note then that whether or not we get $+1$ or -1 for $(((p-1)/2)!)^2$ depends on whether $(p-1)/2$ is *odd* or *even*. If $(p-1)/2$ is odd, it is of the form $2m + 1$, which would make p of the form $4m + 3$. If $(p-1)/2$ is even, it is of the form $2n$, so that p is of the form $4n + 1$. Thus the whole affair reduces to whether or not p is of the form $4n + 3$ or $4n + 1$.

We are now ready to try our hand at proving the general result. This is

Theorem 2.10. If p is a prime of the form $4n + 1$, then we can find an integer x such that $x^2 \equiv -1 \bmod p$. In fact $x = 1 \cdot 2 \cdots ((p-1)/2) = ((p-1)/2)!$ will do the trick.

Proof.   By Wilson's theorem, $(p-1)! \equiv -1 \bmod p$. We write $(p-1)!$ as the product of two terms, the first being the product of all numbers from 1 to $(p-1)/2$, and the second the product of all numbers from $(p+1)/2$ to $p-1$. Now $p-1 \equiv -1 \bmod p$, $p-2 \equiv -2 \bmod p, \ldots, (p+1)/2 \equiv -(p-1)/2 \bmod p$; therefore

$$\left(\frac{p+1}{2}\right) \cdots (p-1) \equiv (-1)(-2) \cdots \left(-\left(\frac{p-1}{2}\right)\right)$$

$$\equiv +\left(\frac{p-1}{2}\right)! \bmod p$$

The plus sign comes from the fact that in the product $(-1)(-2) \cdots (-(p-1)/2)$ there are $(p-1)/2$ terms; since $p = 4n+1$, $(p-1)/2 = 2n$ is *even*. If we let $x = ((p-1)/2)!$, we see that $-1 \equiv (p-1! \equiv x^2 \bmod p$. This is precisely what we have claimed in the statement of the theorem.

What can be said about primes of the form $4n+3$ along the lines of Theorem 2.10? We prove

Lemma 2.7.   If p is a prime of the form $4n+3$, then it is impossible to find an integer x such that $x^2 \equiv -1 \bmod p$.

Proof.   Let p be a prime of the form $p = 4n+3$; then $(p-1)/2 = 2n+1$, an odd number. Suppose that we could find an integer x such that $x^2 \equiv -1 \bmod p$. Then clearly $p \nmid x$. By Fermat's little theorem (Theorem 2.8), $x^{p-1} \equiv 1 \bmod p$. But $x^{p-1} = (x^2)^{(p-1)/2} = (x^2)^{2n+1}$. Going modulo p we get $1 \equiv x^{p-1} \equiv (x^2)^{2n+1} \equiv (-1)^{2n+1} \equiv -1 \bmod p$; this yields that $2 \equiv 0 \bmod p$; that is, $p|2$. Since p is of the form $4n+3$ it is not 2. Thus the assumption of a solution for $x^2 \equiv -1 \bmod p$ led us to a contradiction. The net result is that there is no such x.

Theorem 2.10 and Lemma 2.7 combine to tell us that modulo an odd prime p, -1 is a square if and only if p is of the form $4n+1$. Later we shall deduce from this the beautiful theorem due to Fermat that every prime of the form $4n+1$ is a sum of two squares of integers.

We make an immediate application of these results to finish a story we started earlier. Recall that by modifying Euclid's proof of the existence of an infinite number of primes we were able to prove the stronger statement that there is an infinite number of primes of the form $4n+3$. Earlier, we were in no position to handle $4n+1$; finally we are able to do it.

Theorem 2.11.   There is an infinite number of primes of the form $4n+1$.

Proof.   Suppose that the result is false; then there would be a finite number of primes of the form $4n + 1$. We enumerate these as p_1, $p_2, \ldots, p_k$. Let $a = p_1 p_2 \cdots p_k$ and consider $b = 4a^2 + 1$. Now b is not divisible by any prime of the form $4n + 1$; it furthermore is odd so it is not divisible by 2. But it must be divisible by some prime q; the only possibility is that q is of the form $4n + 3$. Since $q \mid b$ we get $q \mid (4a^2 + 1)$, hence $4a^2 + 1 \equiv 0 \bmod q$. This says that -1 is a *square* modulo q. Invoking Lemma 2.7, we see that for $q = 4n + 3$, -1 is *not* a square modulo q. This contradiction proves our present result.

Exercises

1. Prove that if $a \equiv b \bmod n$ and $c \equiv d \bmod n$, then $a + c \equiv b + d \bmod n$.

2. Let $n > 1$ be an integer and let A_n be the set of all congruence classes $[a] \bmod n$ with a prime to n.
   (a) If $[a], [b] \in A_n$, prove that $[a][b] = [ab]$ is also in A_n.
   (b) Without recourse to Lemma 2.6, prove that if $[a] \in A_n$, there is a $[b] \in A_n$ such that $[a][b] = [1]$. (Hint: Use the Pigeon-Hole Principle.)

3. Let A_n be as in Exercise 2 and let $\phi(n)$ denote the number of elements in A_n. This function is called the Euler ϕ-function. Prove that:
   (a) If $[a] \in A_n$ then $[a]^{\phi(n)} = [1]$.
   (b) If $(a, n) = 1$, then $a^{\phi(n)} \equiv 1 \bmod n$.
   (c) If n is a prime, then $\phi(n) = n - 1$.
   (d) Show how these results imply the little Fermat theorem.

4. If $[a] \in A_n$, let the *order* of $[a]$ be the smallest positive integer k such that $[a]^k = [1]$; write the order of a as $t(a)$. Prove that if $[a]^m = 1$, then $t(a) \mid m$. (Hint: Use the Euclidean algorithm for $t(a)$ and m.)

5. Find the orders of the elements $[1], [2], \ldots, [6]$ in Z_7. (See the preceding exercise for the definition of *order*.)

6. Consider, for p an odd prime, the nonzero elements $[1], [2], \ldots$, $[p - 1]$ in Z_p. (This is the A_p of Exercise 2 for $n = p$.) In Z_p we call $[a]$ a *quadratic residue* if $[a] = [x]^2$ for some integer x: otherwise $[a]$ is called a *quadratic nonresidue*. Prove:
   (a) The product of two quadratic residues is a quadratic residue.
   (b) The product of a quadratic residue and a quadratic nonresidue is a quadratic nonresidue.
   (c) The number of quadratic residues is $(p - 1)/2$, as is the number of quadratic nonresidues.
   (d) The product of two quadratic nonresidues is a quadratic residue.

7. Prove, using the little Fermat theorem, that if $[a]$ is a quadratic residue in Z_p, then $[a]^{(p-1)/2} = [1]$.

8. For $p = 17$ evaluate $((17 - 1)/2)!$ and verify directly the assertion of Theorem 2.10 that $(8!)^2 \equiv -1 \bmod 17$.

9. For $p = 19$ evaluate $((19 - 1)/2)!$ modulo 19 and show that $(9!)^2 \equiv 1 \bmod 19$.

10. If p is a prime of the form $4n + 3$, show that $(((p - 1)/2)!)^2 \equiv 1 \bmod p$.

11. (a) Let p be a prime of the form $4n + 3$ and let $t = ((p - 1)/2)!$.
    (b) Find two primes of the form $4n + 3$ such that $t \equiv 1 \bmod p$; also find two with $t \equiv -1 \bmod p$.

12. Prove the converse of Wilson's theorem: If p divides $1 + (p - 1)!$, then p is a prime.

13. Consider the system consisting of $\square$ and *, with addition and multiplication defined on p. 47. Can you identify this with Z_n for some n? Give the identification explicitly.

7. Applications of the Pigeon-Hole Principle

In considering congruences we enunciated and used a very easy mathematical principle, the so-called Pigeon-Hole Principle. We wish to apply this idea, again in the context of number theory, to derive a classic theorem of Fermat about primes of the form $4n + 1$.

The idea involved in this pigeon-hole notion has applications in many diverse parts of mathematics and, in fact, in many arguments outside mathematics. Some humble uses of it are immediate. In a town of 500 people (we could get by with 367 to account for leap years) there are two people having the same birthday. In New York City, there are two people with the same number of hairs on their heads (if a person has at most 8 million hairs on his head, which seems likely). There are other applications that do not seem mathematical but are. For instance, to describe the motion of a billiard ball one can exploit the idea to great effect. (For this, see page 378 of *Introduction to the Theory of Numbers* 4th ed. by Hardy and Wright (Oxford: Clarendon Press); reference is made there to a "reflected ray.")

We start with an old problem known as the *Chinese Remainder theorem;* before going to this problem in its general statement we consider a special case.

The problem is so named since it goes back to the Chinese poet and mathematician Sun-Tsŭ, about the first century A.D. He asked for a number giving remainders 2, 3, 2 on division by 3, 5, 7, respectively. The problem seems to have arisen at about the same time with the Greeks (Nicomachus, circa 100 A.D.), and later with the Hindus (circa sixth century).

Can we find a positive integer x whose remainder on division by 12 is 5 and on division by 17 is 2? By trial and error we see that the answer is yes; 53 certainly will do. Suppose we ignore this for the moment and ask how we could find this number systematically. The obvious way to begin is to list the positive numbers which on division by 12 leave a remainder of 5. These are 5, 17, 29, 41, 53, 65, ... With these we have at least fulfilled the first requirement. Now we look at the numbers listed and take their remainders on division by 17. By some argument we wish to say that somewhere in this list there is an integer whose remainder on division by 17 is 2. In fact, we should like to specify this "somewhere" a little more precisely. So we consider the following slice of the numbers we listed:

$$5, 5 + 1 \cdot 12, 5 + 2 \cdot 12, \ldots, 5 + 16 \cdot 12$$

Here we have written 17 integers; if we could show that these 17 integers are distinct modulo 17 what could we conclude? Since there are only 17 distinct remainders possible on division by 17, every remainder would be realized. Hence there would be an integer in this list having any preassigned remainder on division by 17, and so, there would certainly be some integer giving 2 as remainder on division by 17. The "somewhere" mentioned earlier would then be described as one of the numbers listed in 5, 5 $+ 1 \cdot 12, \ldots, 5 + 16 \cdot 12$. So everything boils down to showing that these 17 integers give rise to 17 distinct remainders modulo 17. Since the argument in the general case is exactly the one we would also use for 17 we now go to the general case, imitating there the idea we have just proposed.

Theorem 2.12. *(Chinese Remainder theorem).* Let m, n be two positive integers that are relatively prime. Given any two integers a and b, we can find an integer x such that $x \equiv a$ mod m and $x \equiv b$ mod n.

Proof.   Consider the numbers

$$a, a + m, a + 2m, \ldots, a + (n - 1)m$$

They are all congruent to a modulo m. We now consider these integers modulo n; we claim that no two of them are congruent to each other modulo n. If the contrary is true, we would have for some i, j where $0 \leq i < j \leq n - 1$, $a + im \equiv a + jm$ mod n. This leads to $im \equiv jm$ mod n; since m is relatively prime to n, by Lemma 2.5 we get $i \equiv j$ mod n and hence $n|(j - i)$. However, from $0 < j - i < n$ this is manifestly false. Therefore we conclude that the numbers a, $a + m, \ldots, a + (n - 1)m$ give us *distinct* remainders on division by n. Since there are n of these integers having n distinct remainders modulo n, and since there are only n possible remainders modulo n, by the Pigeon-Hole Principle we conclude that every remainder is realized. In particular, we can find an integer w such that $0 \leq w \leq$

$n - 1$ and $a + wn \equiv b$ mod n. The number $a + wn$ is the desired x.

If m and n are not relatively prime, one cannot assert the same conclusion. For instance, it is impossible to find an integer x such that $x \equiv 2$ mod 6 and $x \equiv 1$ mod 8. (Why?)

Theorem 2.12 can be generalized as follows: Let $m_1, \ldots, m_k$ be integers such that any two of them are relatively prime, and let $a_1, \ldots, a_k$ be any integers; then we can find an integer x such that $x \equiv a_1$ mod m_1, $x \equiv a_2$ mod $m_2, \ldots, x \equiv a_k$ mod m_k. We leave the proof as an exercise. (See Exercise 1.)

We now turn to the earlier mentioned goal, namely, a certain theorem of Fermat. First we need a subsidiary lemma.

Lemma 2.8. Let $n > 1$ be an integer; given an integer a we can find two integers x and y, not both 0, whose size is in the range from 0 to $\sqrt{n}$ and that satisfy $xa \equiv y$ mod n.

Proof.   Let m be the largest integer less than or equal to $\sqrt{n}$; that is, $m \leq \sqrt{n} < m + 1$. We consider all numbers of the form $xa - y$ where x can be any of $0, 1, 2, \ldots, m$ and y can be any of $0, 1, 2, \ldots, m$. How many integers do we get this way? Since there are $m + 1$ choices for x, and $m + 1$ independent choices for y, we have a list of $(m + 1)^2$ integers of the form $xa - y$. But since $\sqrt{n} < m + 1$ we get $(\sqrt{n})^2 < (m + 1)^2$; that is, $n < (m + 1)^2$. Having more integers than n, two of these must give the same remainder on division by n. Therefore we can find two numbers $x_1 a - y_1$, $x_2 a - y_2$ such that $x_1 a - y_1 \equiv x_2 a - y_2$ mod n, where $0 \leq x_1 \leq \sqrt{n}$, $0 \leq x_2 \leq \sqrt{n}$, $0 \leq y_1 \leq \sqrt{n}$, $0 \leq y_2 \leq \sqrt{n}$ and where it is not the case that both $x_1 = x_2$ and $y_1 = y_2$. Let $x = x_1 - x_2$, $y = y_1 - y_2$; then x and y are not both 0, and in size they lie in the range from 0 to $\sqrt{n}$. But $x_1 a - y_1 \equiv x_2 a - y_2$ mod n yields $(x_1 - x_2)a \equiv y_1 - y_2$ mod n; that is, $xa \equiv y$ mod n. We have produced the integers x and y required in the lemma.

We are now in a position to deliver our promise to prove the result due to Fermat mentioned earlier.

Theorem 2.13.   (Fermat) Let p be a prime of the form $4n + 1$. Then we can find integers x and y such that $x^2 + y^2 = p$.

Proof.   By Theorem 2.10 we have that when p is of the form $4n + 1$ then $a = ((p - 1)/2)!$ satisfies $a^2 \equiv -1$ mod p.

By Lemma 2.8 we can find integers x and y, not both 0, such that x and y are smaller in size than $\sqrt{p}$ (we can say *smaller* since $\sqrt{p}$

is not an integer), and such that $xa \equiv y$ mod p. Since x and y are not both 0 we know that $x^2 + y^2 \neq 0$. Since the size of x is less than $\sqrt{p}$, $x^2 < p$; similarly $y^2 < p$. Hence $0 < x^2 + y^2 < 2p$.

Since $xa \equiv y$ mod p, $(xa)^2 \equiv y^2$ mod p, which implies that $x^2a^2 \equiv y^2$ mod p. But $a^2 \equiv -1$ mod p; hence $-x^2 \equiv y^2$ mod p. This yields that $x^2 + y^2 \equiv 0$ mod p and so $p|(x^2 + y^2)$. Now look at this! $x^2 + y^2$ is *positive*, is *smaller* than $2p$, and is a *multiple* of p. What can it be? It can only be p. Thus $x^2 + y^2 = p$. We have proved the theorem.

Actually, the proof gives a constructive, albeit not efficient, method for finding x and y. We illustrate with $p = 29$. Here $a = 17$. The proof suggests that we should look for repeats, mod 29, in the numbers $xa - y$ as x and y range from 0 to m; here m is defined by $m < \sqrt{29} < m + 1$; that is, $m = 5$. We make a table of $xa - y$ mod 29.

	$x = 0$	$x = 1$	$x = 2$	$x = 3$	$x = 4$	$x = 5$
$y = 0$	0	17	5	22	10	27
$y = 1$	28	16	4	21	9	26
$y = 2$	27	15	3	20	8	25
$y = 3$	26	14	2	19	7	24
$y = 4$	25	13	1	18	6	23
$y = 5$	24	12	0	17	5	22

We see that we have repeats: For instance, 0 occurs as $0a - 0$ and as $2a - 5$. So we set $x_1 = 0$, $y_1 = 0$, $x_2 = 2$, $y_2 = 5$. Then, according to the proof, $x = x_1 - x_2 = -2$ and $y = y_1 - y_2 = -5$ gives us $x^2 + y^2 = 29$ (which indeed they do). We have other repeats. For example, 22 occurs as $3a - 0$ and as $5a - 5$; hence we could use $x_1 = 3$, $y_1 = 0$, $x_2 = 5$, $y_2 = 5$, and get $x = x_1 - x_2 = 3 - 5 = -2$ and $y = y_1 - y_2 = 0 - 5 = -5$, yielding $x^2 + y^2 = 29$. In fact we could use any of the repeats to come up with an answer.

Note that if p is a prime of the form $4n + 3$ then we *cannot* find integers x and y such that $x^2 + y^2 = p$. A simple way to see this is as follows: If we go modulo 4, then $x \equiv 0, 1, 2$, or 3 mod 4, so that $x^2 \equiv 0, 1$, $2^2 \equiv 0$, or $3^2 \equiv 1$ mod 4. Therefore $x^2 \equiv 0$ or 1 mod 4. Similarly $y^2 \equiv 0$ or 1 mod 4. Hence $x^2 + y^2 \equiv 0, 1$, or 2 mod 4, showing that $x^2 + y^2 \not\equiv 3$ mod 4. In short, $x^2 + y^2$ cannot be of the form $4m + 3$; in particular, $p = 4n + 3$ is not so realizable.

Returning to Fermat's theorem, a natural question presents itself: In how many distinct ways can a given prime p of the form $4n + 1$ be written as a sum of two squares? In other words, if $p = x^2 + y^2$, x and y integers, how many choices do we have for x and y? Consider $p = 17$; $17 = 4^2 + 1^2$ so here $x = 4$, $y = 1$ is such a choice. But $17 = 1^2 + 4^2$ so $x = 1$, $y = 4$

is another solution. Furthermore we can adorn 4 or 1 or both with a minus sign, so we get still other solutions. We would like to consider all of these as essentially the same, that is, interchange of x and y or change of sign shall be considered irrelevant. We shall now show that except for such changes the representation of p as a sum of two squares is unique.

From what we said above, the problem of the uniqueness of the expression as a sum of two squares reduces to the following: Let p be a prime of the form $4n + 1$ and suppose $p = x^2 + y^2 = a^2 + b^2$, where x, y, a, b are positive integers and where $x > y$, $a > b$. Our aim is to show that $x = a$ and $y = b$.

We do some algebra. Consider

$$(xa + yb)(xb + ya) = x^2ab + y^2ab + b^2xy + a^2xy = (x^2 + y^2)ab$$
$$+ (b^2 + a^2)xy = pab + pxy = p(ab + xy)$$

Thus $p|(xa + yb)(xb + ya)$, and so $p|(xa + yb)$ or $p|(xb + ya)$.
Suppose $p|(xb + ya)$. Now

$$(xb + ya)^2 + (xa - yb)^2 = x^2b^2 + y^2a^2 + 2xyab + x^2a^2 + y^2b^2$$
$$- 2xyab = x^2(a^2 + b^2) + y^2(a^2 + b^2)$$
$$= (x^2 + y^2)(a^2 + b^2) = p^2$$

Since p divides $xb + ya$, and $xb + ya$ is positive, $xb + ya$ is a positive multiple of p; we deduce $(xb + ya)^2 \geqq p^2$. On the other hand, from $(xb + ya)^2 + (xa - yb)^2 = p^2$ we get $(xb + ya)^2 \leqq p^2$. Hence $(xb + ya)^2 = p^2$, from which we get $xa - yb = 0$; that is, $xa = yb$. Since $x > y$ and $a > b$ this is impossible.

So we are left with the other possibility, namely $p|(xa + yb)$. But then $(xa + yb)^2 + (xb - ya)^2 = p^2$ by a calculation similar to the one made above. Since $xa + yb$ is positive and divisible by p we get, as above, $xb = ya$ and $(xa + yb)^2 = p^2$. Since $xa + yb$ is positive, $xa + yb = p$. Thus $xab + yb^2 = (xa + yb)b = pb$; since $xb = ya$, $xab = xba = (ya)a = ya^2$, hence $pb = xab + yb^2 = ya^2 + yb^2 = y(a^2 + b^2) = py$. Because $pb = py$ we get $b = y$. Then $xb = ya$ tells us that $x = a$. This proves the uniqueness claimed.

We have seen that certain primes are expressible as a sum of two squares of integers and others are not. One can then ask: Fine, two squares do not work, is there some fixed number of squares that will? The question seems to have been first raised by Diophantus and was settled, in 1770, by Lagrange. He proved that every positive integer is a sum of four squares. In other words, given an integer $n > 0$ then $n = a^2 + b^2 + c^2 + d^2$ for some integers a, b, c, d. We shall use this result in Chapter 5.

Incidentally, it is easy to see that four is the minimum number of squares that work. To see this one merely notes that 7, for instance, is not a sum of three squares of integers.

Diophantus was a Greek mathematician who lived in Alexandria during the third century A.D. His great work was called the *Arithmetica*. Apparently it consisted of thirteen books, only six of which have survived. It is in essence a large collection of examples of polynomial equations, the problem being to find integral or rational solutions. Diophantus was always content to find one solution rather than trying to find all solutions, as we usually do nowadays.

Exercises

1. Prove the more general Chinese Remainder theorem, namely, given integers $m_1, \ldots, m_k$, any two of which are relatively prime to each other, and integers $a_1, \ldots, a_k$, then there exists an integer x such that $x \equiv a_1 \bmod m_1$, $x \equiv a_2 \bmod m_2, \ldots, x \equiv a_k \bmod m_k$.

2. Using the Pigeon-Hole Principle prove that the decimal expansion of any rational number (quotient of two integers) is a repeating decimal (i.e., after some point, the decimal expansion repeats the same block). (Example: $2479/3300 = 0.75121212\ldots$)

3. Let A be the set of all integers that can be written as a sum of two squares of integers. Prove that if a and b are in A, then so is ab.

4. Let $n = p_1^{\alpha_1} p_2^{\alpha_2} \cdots p_k^{\alpha_k}$ be the factorization of n as the product of powers of the distinct primes $p_1, \ldots, p_k$. Prove that n is a sum of two squares if and only if for every p_i, which is of the form $4m + 3$, the corresponding α_i is even.

5. Prove that no integer of the form $8n + 7$ is a sum of three squares. (Hint: Look at possible squares mod 8.)

8. Waring's Problem

It was mentioned at the end of the preceding section that every positive integer can be written as a sum of four squares. In this section we shall briefly discuss the analogous problems for cubes, fourth powers, and higher powers.

How many cubes do we need in order to write every number as a sum of cubes? (Note: In this section, except for Exercise 4, we shall be talking only about positive integers.) Certainly at least 7, for the only way of writing 7 as a sum of cubes is to write it as a sum of seven 1's. When we come to the number $8 = 2^3$, it is itself a cube, so one is all we need. The number $9 = 8 + 1$ is the sum of two cubes. Further along we reach the number 15; our most economical procedure is to write it as a sum of 8 and seven 1's, thus using eight cubes.

A pattern is beginning to appear. For the number 7 and the number 15 we had to use seven 1's, and for 15 we used an 8 as well. The next number of this kind is 23; for it we must again use seven 1's together with two 8's this time, for a total of nine cubes. But at the next number 31 the pattern breaks off. For 31 we need not use seven 1's and three 8's; instead we can use 27 and four 1's.

To summarize: It takes nine cubes to write 23 as a sum of cubes. Does any number need even more? We shall return to this shortly.

But first let us dip back in mathematical history to the year 1770 when the English mathematician Edward Waring made the following remarkable statement: Any number is a sum of 4 squares, 9 cubes, 19 fourth powers, and so on.

Edward Waring (1734–1798) began his career as a doctor. He gave up his practice in 1770 and became the Lucasian professor of Mathematics at Cambridge University. He wrote numerous books and made several contributions to number theory and to algebra.

Some people think he only meant to say that for every n there is *some* number such that every integer is expressible as a sum of that many nth powers. Even this much turned out to be hard to prove; it wasn't proved until David Hilbert (the greatest mathematician of his time) published a proof in 1909.

But where did Waring get the particular numbers 4, 9, and 19 for squares, cubes, and fourth powers? And did he have some idea as to how to continue? It seems quite possible that he did, for we can arrive at the guess 19 by a little analysis such as we did above for cubes. This time we decide to stay below the number $3^4 = 81$, for as soon as 81 is available the situation gets quite complex.

We examine the numbers 15, 31, 47, . . . which can only be expressed as a sum of fourth powers as follows: Use fifteen 1's together with the appropriate number of 16's. The largest number of this type, lying below 81, is 79. Another way of locating 79 is as follows: Take 80, the largest multiple of 16 lying below 81, and then subtract 1. Now in expressing 79 as a sum of fourth powers, the best we can do is:

79 = fifteen 1's plus four 16's

for a total of 19 fourth powers. So there is Waring's 19. Notice that our little investigation says nothing about the much harder question as to whether every number is really a sum of 19 fourth powers; we only know that no number smaller than 19 can possibly work. We shall return to this shortly.

Let us try fifth powers. We stay below $3^5 = 243$. The highest multiple of $2^5 = 32$ below 243 is $7 \cdot 32 = 224$. Subtract 1 to get 223. The most economical expression for 223 is:

223 = thirty-one 1's plus six 32's

So the number 223 needs 37 fifth powers. Perhaps Waring would have given 37 as the next number on his list (if somebody had asked him).

We can work out a formula for the general case of the numbers 4, 9, 19, 37, ... Divide 3^n by 2^n using the Euclidean algorithm. Say the quotient is q and the remainder r, so that

$$3^n = 2^n q + r$$

Notice that 3^n is not divisible by 2^n, so that r is not 0. Thus we have $0 < r < 2^n$. The number $2^n q$ is the largest multiple of 2^n that lies below 3^n. We take $s = 2^n q - 1$. Then the most economical expression for s as a sum of nth powers is to write it as a sum of $2^n - 1$ ones and $q - 1$ 2^n's, thus using up a total of $(2^n - 1) + (q - 1) = 2^n + q - 2$ nth powers. This is the formula we are looking for. Let us call this number h, and let us write it as $h(n)$, since h depends on n. Of course q also depends on n, so we shall now be more careful and write $q(n)$. All in all we have

$$h(n) = 2^n + q(n) - 2$$

and what we know is this: *There is a number that needs $h(n)$ nth powers in order to write it as a sum of nth powers.*

Let us do the cases $n = 2$ to $n = 5$ again in the present notation, and add the next two cases ($n = 6$ and 7).

$n = 2$. $3^2 = 9$, $2^2 = 4$, $9 = 2 \cdot 4 + 1$, $q = 2$, $h(2) = 2^2 + 2 - 2 = 4$

$n = 3$. $3^3 = 27$, $2^3 = 8$, $27 = 3 \cdot 8 + 3$, $q = 3$, $h(3) = 2^3 + 3 - 2 = 9$

$n = 4$. $3^4 = 81$, $2^4 = 16$, $81 = 5 \cdot 16 + 1$, $q = 5$, $h(4) = 2^4 + 5 - 2 = 19$

$n = 5$. $3^5 = 243$, $2^5 = 32$, $243 = 7 \cdot 32 + 19$, $q = 7$, $h(5) = 2^5 + 7 - 2 = 37$

$n = 6$. $3^6 = 729$, $2^6 = 64$, $729 = 11 \cdot 64 + 25$, $q = 11$, $h(6) = 2^6 + 11 - 2 = 73$

$n = 7$. $3^7 = 2187$, $2^7 = 128$, $2187 = 17 \cdot 128 + 11$, $h(7) = 2^7 + 17 - 2 = 143$

We shall conclude this section by stating the known facts about whether any number is expressible as a sum of $h(n)$ nth powers. Except for a little nibble for $n = 4$ in Exercise 5, not a shred of proof will be offered; the proofs are much too long and difficult for this book.

First let us introduce the notation that is universally used in discussing Waring's problem. Write $g(n)$ for the smallest integer for which it is true that any number can be written as a sum of $g(n)$ nth powers. Notice that we showed above that $g(n) \geqq h(n)$, and now the question is: Are the two actually equal? We discuss first the low values of n, one at a time.

$n = 2$. We have $h(2) = 4$, and $g(2)$ is also 4. This is Lagrange's theorem asserting that every number is a sum of four squares, discussed in the preceding section.

$n = 3$. We have $h(3) = 9$. $g(3)$ is also 9, as was proved by A. Wieferich in 1909 (with a gap that was filled by A. J. Kempner in 1912).

$n = 4$. We have $h(4) = 19$. Liouville was the first to get some bound for $g(4)$. He found $g(4) \leq 53$, and his method is followed in Exercise 5. This was whittled down through the years until Dickson achieved $g(4) \leq 35$ in 1933. There was no further progress for 28 years. Then, in 1971, François Dress lowered this to $g(4) \leq 30$, and in 1972 H. E. Thomas, Jr. achieved $g(4) \leq 23$. Both Dress and Thomas made use of large computers.

Joseph Liouville was born in Saint Omer, France, in 1809, and died in Paris in 1882. In addition to his numerous contributions to virtually all branches of mathematics, he was important as a mathematical "statesman." In 1836 he founded the *Journal des Mathématiques Pures et Appliqués;* he edited it for forty years. It was in this journal that Liouville published Galois's remarkable work.

Leonard Eugene Dickson (1874–1954) dominated algebra and number theory in the United States for an entire generation. After receiving the first Ph.D. awarded by the newly founded University of Chicago in 1896, he studied in Europe, and then returned in 1900 to a professorship at Chicago for the rest of his career. He wrote 270 research papers and 18 books, including the monumental three-volume *History of the Theory of Numbers.* An instructorship in mathematics at Chicago now bears his name.

$n \geq 5$. Starting at $n = 5$, the problem "settles down." Difficult and ingenious work by Chen, Dickson, Niven, Pillai, and Vinogradov has virtually solved the problem of finding $g(n)$ for $n \geq 5$. Let us recall the numbers q and r we introduced, satisfying $3^n = q2^n + r$, and let us now acknowledge that r also depends upon n by writing $r(n)$. It makes a crucial difference whether or not the following is true:

(1) $r(n) \leq 2^n - q(n)$

If (1) is true, then $q(n) = h(n)$ for $n \geq 5$. If (1) is not true, then $g(n)$ is slightly larger than $h(n)$; it can still be written explicitly but we shall not bother to do so. However, there is the following provocative thought in addition: It is possible that (1) always holds. This has been checked all the way up to $n = 200,000$. So, for $5 \leq n \leq 200,000$, we do know that $g(n) = h(n) = 2^n + q(n) - 2$.*

*Here are two references where you can read more about Waring's problem: Shanks, *Solved and Unsolved Problems of Number Theory*, p. 211; Hardy and Wright, *An Introduction to the Theory of Numbers*, ch. 21.

The article "Waring's problem" by W. J. Ellison in the *American Mathematical Monthly* vol. 78 (1971), pp. 10–36, is an advanced scholarly summary, including a large bibliography. In the December 1973 *Scientific American*, Martin Gardner's column, "Mathematical Games," is devoted to Waring's problem.

88 *Number Theory*

Exercises*

1. Verify that, with the sole exception of 23, every integer less than 100 is a sum of eight cubes. (Remark: In 1939 Dickson proved that all integers other than 23 and 239 are sums of eight cubes.)

2. Prove that no integer of the form $9n + 4$ or $9n + 5$ can be written as a sum of three cubes. (Hint: Check out the possibilities for a cube modulo 9, and then see what you can get when adding three cubes.)

   (Notice that this implies that an infinite number of integers need at least four cubes. It is not known whether the same statement can be made with five instead of four; in other words, it is conceivable that from some point on in the list of integers, four cubes are sufficient. The best that has been proved for sure is a theorem of Linnik asserting that, from some point on, seven cubes are enough.)

4. In this exercise, negative integers are allowed throughout.

   (a) Prove that for any integer n, $n^3 - n$ is divisible by 6. (We are repeating this from Exercise 12 in Section 1.)

   (b) Check the identity

   $$n^3 - 6x = n^3 - (x + 1)^3 - (x - 1)^3 + 2x^3$$

   (c) Use parts (a) and (b) to prove that any integer is a sum of five cubes.

5. (a) Verify the identity

   $$\begin{aligned}
   6(a^2 + b^2 + c^2 + d^2)^2 = {}& (a + b)^4 + (a + c)^4 + (a + d)^4 \\
   & + (b + c)^4 + (b + d)^4 + (c + d)^4 \\
   & + (a - b)^4 + (a - c)^4 + (a - d)^4 \\
   & + (b - c)^4 + (b - d)^4 + (c - d)^4
   \end{aligned}$$

   (b) Assuming Lagrange's theorem that any integer is a sum of four squares, deduce from part (a) that any integer of the form $6x^2$ is a sum of 12 fourth powers.

   (c) Using part (b) and Lagrange's theorem again, prove that any integer is a sum of 53 fourth powers. (Hint: Put the given integer in the form $6q + r$ with $r \leqq 5$.)

6. Prove that any number of the form $16n + 15$ needs at least 15 fourth powers when written as a sum of fourth powers. (Hint: Look at fourth powers mod 16.)

---

*In all exercises except 4 the integers in question are to be taken as positive.

9. Fermat and Mersenne Primes

In this final section of our chapter on number theory we return to prime numbers, and we launch a search for them among numbers of the form $2^n \pm 1$ (a power of two increased or diminished by one).

Why should numbers of this particular form get special attention? One answer is that the numbers $2^n \pm 1$ have such a charmingly simple form that for this reason alone most people would be interested in discovering which of them are primes. But actually there are other reasons—partly historical, partly because of connections with other parts of mathematics—that give the subject an independent importance. We shall discuss some of these reasons at the end of the section.

Let us try out the low cases. There is not much point in starting below 4. In the table below we assemble the facts for the powers of 2 from 2^2 to 2^8 (4 to 256). We notice several things. The numbers in question are not all prime, but there is a good sprinkling of primes among them. There is one case, namely 64, where both neighbors (63 and 65) are composite. After the pair 3, 5 surrounding 4, there is no further case in the table where both neighbors are prime. This last observation is not an accident. After all, given the three consecutive integers $2^n - 1$, 2^n, and $2^n + 1$, one of them must be divisible by 3, and certainly 2^n is not that one.

n	2^n	$2^n - 1$	$2^n + 1$
2	4	3 prime	5 prime
3	8	7 prime	$9 = 3 \cdot 3$
4	16	$15 = 3 \cdot 5$	17 prime
5	32	31 prime	$33 = 3 \cdot 11$
6	64	$63 = 3^2 \cdot 7$	$65 = 5 \cdot 13$
7	128	127 prime	$129 = 3 \cdot 43$
8	256	$255 = 3 \cdot 5 \cdot 17$	257 prime

To get deeper into the subject, we pause to learn a piece of easy algebra.

Lemma 2.9. For any positive integer r, the polynomial $x^r - 1$ is divisible by $x - 1$. For odd r, $x^r + 1$ is divisible by $x + 1$.

Proof. After the successful experiments

$$x^2 - 1 = (x - 1)(x + 1)$$
$$x^3 - 1 = (x - 1)(x^2 + x + 1)$$
$$x^4 - 1 = (x - 1)(x^3 + x^2 + x + 1)$$

it is not hard to see that we can factor $x^r - 1$ explicitly:

(2) $x^r - 1 = (x - 1)(x^{r-1} + x^{r-2} + \cdots + x + 1)$

For odd r, experimentation leads to

$$x^3 + 1 = (x + 1)(x^2 - x + 1)$$
$$x^5 + 1 = (x + 1)(x^4 - x^3 + x^2 - x + 1)$$

Then for a general odd r, say $r = 2m + 1$, we venture to write

(3) $x^{2m+1} + 1 = (x + 1)(x^{2m} - x^{2m-1} + \cdots + x^2 - x + 1)$

Equation (3), like Equation (2), can be verified by simply multiplying the right-hand side.

For completeness we remark that there is still another identity of this kind:

$$x^{2m} - 1 = (x + 1)(x^{2m-1} - x^{2m} + \cdots - x^2 + x - 1)$$

showing that $x^r - 1$ is divisible by $x + 1$ if r is even.

All of these facts are special cases of a result called the *factor theorem* which asserts the following: If $f(x)$ is a polynomial with integral coefficients that vanishes when x is set equal to an integer a (in other words, $f(a) = 0$), then $f(x) = (x - a)\,g(x)$ where $g(x)$ is again a polynomial with integral coefficients.

With the aid of Lemma 2.9, we can drastically restrict the possible values of n that make $2^n - 1$ or $2^n + 1$ prime. The following theorem does just that.

Theorem 2.14.   If n is composite, then $2^n - 1$ is composite. If $n > 1$ and n is not a power of 2, then $2^n + 1$ is composite.

Proof.   For the first half of the theorem, assume that n is composite and write $n = rs$, with r and s both positive integers greater than 1. Set $x = 2^s$, and we then have $2^n = 2^{rs} = (2^s)^r = x^r$. So $2^n - 1 = x^r - 1$. By Lemma 2.9 $x^r - 1$ is divisible by $x - 1$. Since $s \geq 2$, we have $x = 2^s \geq 4$, $x - 1 \geq 3 > 1$. Since $s < n$ we have $x = 2^s < 2^n$, and $x - 1 < 2^n - 1$. Thus the factor $x - 1$ of $2^n - 1$ is neither 1 nor $2^n - 1$. We have that $2^n - 1$ is composite.

We turn to the second half of the theorem. This time we assume that n is not a power of 2. As a consequence we can write $n = rs$ with r odd and bigger than 1. There is a special possibility we must dispose of: Perhaps n is an odd prime, in which case r will have to be equal to n. But in that case we can apply the second half of Lemma 2.9, with $x = 2$, to see that $2^n + 1$ is divisible by 3. So we can continue the proof under the assumption that neither r nor s is 1. Again we set $x = 2^s$ and find $2^n = x^r$, so that $2^n + 1 = x^r + 1$. Once again we make use of the second half of Lemma 2.9 to find that $x + 1$ divides $2^n + 1$, and it is evidently a proper divisor.

The theorem shows that there is no hope for $2^n - 1$ to be a prime unless n is a prime. This calls for a change of notation. We write $M_p = 2^p - 1$, where p is assumed to be prime. Here the letter M honors Mersenne, and the numbers M_p are called the Mersenne numbers.

> Father Marin Mersenne (1588–1648) joined the Minim friars in 1611 and settled at the convent of L'Annonciade in Paris in 1620. He was active as a theologian as well as a mathematician. He collaborated in some of Descartes' work and carried on a vigorous correspondence with Fermat. In the preface to his *Cogita Physico-Mathematica* (Paris, 1644) he made the remarkable assertion that, for $p \leq 257$, M_p is prime exactly for $p = 2, 3, 5, 7, 13, 17, 19, 31, 67, 127$, and 257. When Seelhoff (1886) proved that M_{61} is prime and Cole (1903) proved that M_{67} is composite, it was suggested that 67 was a misprint for 61. But before long three other errors were found in the Mersenne list: M_{89} is prime (Powers, 1911); M_{107} is prime (Powers and Fauquemberque independently in 1914); and M_{257} is composite (Kraitchik, 1922, verified by Lehmer, 1933).

M_p is indeed a prime in the first four cases: $p = 2, 3, 5, 7$ ($M_p = 3, 7, 31, 127$). But optimistic hopes are dashed at the very next prime 11. We have $M_{11} = 2047$, and this small number is fairly easy to factor into $23 \cdot 89$.

We shall interrupt the discussion of Mersenne primes at this point, with the promise to return to them later. Let us examine the numbers $2^n + 1$. Theorem 2.14 restricts our search for primes to the case where n is a power of 2, and so we write $n = 2^t$, $F_t = 2^{2^t} + 1$. We might as well start our investigation with the rock-bottom value: $t = 0$. The first five F's are:

t	F_t
0	3
1	5
2	17
3	257
4	65,537

The numbers 3, 5, 17, and 257 are seen to be prime at a glance. The number 65,537 cannot be disposed of quite so quickly. But it is not a hard job. Possible prime factors need only be checked up to the square root (see Exercise 1 in Section 1). The square root of 65,537 is a trifle larger than 256, so the largest prime factor that needs to be tested is 251. Furthermore, a big reduction in the work can be made by using Theorem 2.15. Finally, there are reliable tables of primes going up to 10 million. One way or another, we discover that 65,537 is prime.

Why the letter F in the notation $F_t = 2^{2^t} + 1$? You guessed it—
Fermat was at work again. But this was one of his less-inspired moments.
In 1640 he wrote that he was convinced that all the numbers F_t are prime,
but was unable to prove it. Euler was unimpressed. He tackled $F_5 =$
4,294,967,297. In the eighteenth century, factoring a 10-digit number was
not something to be undertaken lightly. But Euler was able to improve his
prospects by proving Theorem 2.15.

> The Swiss mathematician Leonhard Euler (1707–1783) wrote discerningly
> on every branch of the mathematics of his time. He was probably the most
> prolific of all mathematicians. Publication of his collected works remains
> incomplete, although over seventy fat volumes are published. The following
> is quoted from Bell's *Men of Mathematics* (pp. 145–146). "He was very fond
> of children . . . and would often compose his memoirs with a baby in his
> lap while the older children played all about him."

Theorem 2.15.   Let q be a prime factor of F_t. Then $q \equiv 1 \bmod 2^{t+1}$;
that is, q has the form $2^{t+1}u + 1$.

To make the proof of this theorem go a little more smoothly, we shall
separate out the crucial part as a lemma.

Lemma 2.10.   Let q be a prime, let k be a power of 2, and assume that
$2^k \equiv -1 \bmod q$. Then $2k$ divides $q - 1$.

Proof.   The idea of the proof is to use the little Fermat theorem (Theorem
2.8), make various maneuvers modulo q, and take advantage of the
assumption that k is a power of 2.

First, square the given congruence $2^k \equiv -1 \bmod q$. The result is

(4) $2^{2k} \equiv 1 \bmod q$

By Theorem 2.8 we have

(5) $2^{q-1} \equiv 1 \bmod q$

Let m be the greatest common divisor of $2k$ and $q - 1$. We have
$m = a(2k) + b(q - 1)$ for suitable integers a and b. From all this we
can deduce $2^m \equiv 1 \bmod q$. The details are as follows. Raise (4) to
the ath power and (5) to the bth power:

(6) $2^{a(2k)} \equiv 1 \bmod q$
(7) $2^{b(q-1)} \equiv 1 \bmod q$

Multiply (6) and (7):

$2^{a(2k)+b(q-1)} \equiv 1 \bmod q$

So we have $2^m \equiv 1 \bmod q$. Now, recall that k is a power of 2, and
so is $2k$. Since m divides $2k$, it is also a power of 2. We are going to
show that $m = 2k$. Making an indirect proof, we assume $m < 2k$.

Then since m and k are both powers of 2, we have that m divides k. This enables us to raise $2^m \equiv 1$ mod q to an appropriate power to get $2^k \equiv 1$ mod q. But now 2^k is congruent to both 1 and -1 mod q, which is preposterous. This contradiction completes the proof of Lemma 2.10.

With this lemma at hand, the proof of Theorem 2.15 goes swiftly.

Proof of Theorem 2.15.   To shorten the writing, set $k = 2^t$. Then $2^k + 1 = F_t$. Since q divides F_t, we deduce $2^k \equiv -1$ mod q. The hypotheses of Lemma 2.10 are fulfilled and we deduce that $q - 1$ is divisible by $2k$. It only remains for us to notice that $2k = 2 \cdot 2^t = 2^{t+1}$.

Let us return to the Fermat number $F_4 = 2^{16} + 1 = 65,537$ for a moment. We said earlier that it was relatively easy to check that F_4 is prime. Now we can make it even easier. For, by Theorem 2.15, the only prime divisors we need to check are those of the form $32u + 1$. If we discard composite numbers, and stay below the square root of 65,537, we find that only 97 and 193 need to be tested. We might as well complete the job and make our discussion of 65,537 entirely self-contained. By long division, we find

$$65,537 = 97 \cdot 675 + 62$$
$$65,537 = 193 \cdot 339 + 110$$

Hence 65,537 is not divisible by 97 or 193. We have proved in full that 65,537 is prime.

Actually, by using congruences we could even have bypassed long division in favor of a little mental arithmetic. For the case of 193, for instance, the question is whether $65,537 = 2^{16} + 1$ is divisible by 193. In other words, we are asking whether $2^{16} \equiv -1$ mod 193. Assume that this is true. All congruences in this paragraph will be mod 193. Notice that $192 = 3 \cdot 2^6$, so that $3 \cdot 2^6 \equiv -1$. Cube it to get $27 \cdot 2^{18} \equiv -1$. But $27 \cdot 2^{18} = 27 \cdot 4 \cdot 2^{16} = 108 \cdot 2^{16} \equiv -108$. There is our contradiction. We invite the reader to work up a similar discussion for 97.

Now we can join Euler in his attack on F_5. By Theorem 2.15, only prime divisors of the form $64u + 1$ need to be tried. Discarding composites, we find that the first five candidates are

$$193, 257, 449, 577, 641$$

Euler hit the jackpot on the fifth try! $F_5 = 2^{32} + 1$ is divisible by 641. Again this is delightfully simple by congruences. With all congruences taken modulo 641 we have $5 \cdot 2^7 = 640 \equiv -1$, whence (take fourth powers) $625 \cdot 2^{28} \equiv 1$. But $625 \equiv -16 = -2^4$. Hence $2^{32} \equiv -1$, and we are through.

If you would like to see the other factor of F_5, divide and get

$$F_5 = 641 \cdot 6,700,417$$

Euler checked that the second factor is prime.

Nohting much happened for a century until, in 1880, F. Landry factored F_6 into prime factors:

$$F_6 = 2^{64} + 1 = 274,117 \cdot 67,280,421,310,721$$

In 1905 Morehead and Western independently proved that F_7 is composite, but they did not exhibit any factorization. How was that possible?

To answer this question, let us think of possible ways to check whether a given number q is prime. There is of course the poor man's way: Test every possible divisor. At a slightly higher level of sophistication (as we remarked above in factoring 65,537), we only have to check whether q is divisible by some prime less than $\sqrt{q}$. For numbers as big as the ones we are now discussing, that does not help enough. In the case of Fermat numbers, Theorem 2.15 lessens the possibilities, but still not enough.

There is Wilson's theorem (Theorem 2.9) which says that q is prime if and only if q divides $1 + (q - 1)!$. But if q is large, $(q - 1)!$ is astronomical, and this suggestion is truly ludicrous.

How about Fermat's little theorem? In other words, why not try to check whether q is prime by seeing if $2^{q-1} \equiv 1 \bmod q$ is true. Again you might think this is hopeless since 2^{q-1} is going to be so large. However, remember that we are entitled to carry out the computation modulo q. In fact, this makes the computation realistic; we shall discuss this point further below.

There is just one trouble. If q is prime, we have $2^{q-1} \equiv 1 \bmod q$. But no one said anything about the converse. The reason nothing was said is that it is false. The first counter-example is 341; it is composite, being equal to $11 \cdot 31$, and yet $2^{340} \equiv 1 \bmod 341$. (Challenge: Check this by modular mental arithmetic.) So it is perfectly possible for $2^{q-1} - 1$ to be divisible by q even though q is composite. In some sense the phenomenon is rare. But, perversely enough, for the Fermat and Mersenne numbers we are interested in testing, failure is certain (see Exercises 4 and 5). In a way, Exercise 6 is a more spectacular instance of failure of the converse of Fermat's little theorem, for in that example, failure occurs with any number a used in place of 2.

Despite these discouraging examples, there *is* a test of this kind that is effective for Fermat numbers. In formulating it, we change in three ways the idea of using $2^{q-1} \equiv 1 \bmod q$ as a test for the primality of q: We use 3 instead of 2, we divide $q - 1$ by 2, and we replace 1 by -1 on the right side of the congruence. The result is the following theorem, due to Pepin.

Theorem.   Let q be a Fermat number, $q = 2^{2^t} + 1$, with $t > 0$. Then q is prime if and only if $3^{(q-1)/2} \equiv -1 \bmod q$.

As a modest illustration of the theorem, let us take $t = 2$, $q = 17$. To see whether $3^8 \equiv -1 \bmod 17$ holds, we use a little modular arithmetic. With all congruences taken modulo 17, we have $3^3 = 27 \equiv 10$, $3^6 = 10^2 = 100 \equiv -2$, $3^8 \equiv 9(-2) = -18 \equiv -1$. So this checks with the fact that 17 is prime.

Pepin's theorem is a little too difficult for us to prove, and we consequently ask the reader to take it on faith.

Let us see how this might actually be used to test a Fermat number $F_t = q$. We are called on to check whether $3^{(q-1)/2}$ is congruent to -1 modulo q. Note that $(q - 1)/2 = 2^{2^t-1}$. We can in fact reach $3^{(q-1)/2}$ by $2^t - 1$ successive squarings. Let us illustrate with the case $t = 3$, $q = 257$. We have $(3^2)^2 = 3^4$, $(3^4)^2 = 3^8$, $(3^8)^2 = 3^{16}$. We have now squared 3 four successive times and reached $3^{16} = 3^{2^4}$. In the same way, three more squarings brings us to $3^{2^7} = 3^{128}$. Notice again that in checking whether $3^{128} \equiv -1 \pmod{257}$, we need not work out the large number 3^{128}, for we can work modulo 257.

Likewise, in the general case, the proposed computation will go as follows: Start with the number 3 and perform $2^t - 1$ successive squarings. After each squaring reduce mod $q = F_t = 2^{2^t} + 1$. If the final result is -1, F_t is prime; if not, F_t is composite (but note that in the composite case no factor of F_t will have been discovered).

For $t = 7$, the procedure calls for $2^7 - 1 = 127$ squarings followed by reduction modulo $2^{128} + 1$. Now $2^{128} + 1$ is a number of 39 digits. So what it all boils down to is this: We must take a number having perhaps as many as 39 digits, square it, divide by a number of 39 digits, and find the remainder. Repeat this 127 times. As long ago as 1905, Morehead and Western were able to carry this out, and they found $2^{128} + 1$ to be composite. On today's electronic computers, as we shall mention below, much bigger numbers have been handled this way.

Sixty-five years after Morehead and Western demolished F_7, Brillhart and Morrison succeeded in factoring F_7. The work was done at UCLA on an IBM 360/91 (a very large computer). The result is worth recording as perhaps the most spectacular factorization ever accomplished:

$$F_7 = 2^{128} + 1 = 59649589127497217 \cdot 5704689200685129054721$$

For the next number F_8 a challenge remains. In 1909 Morehead and Western (jointly) found that F_8 is composite, but its factors are unknown. The lowest Fermat number whose status as a prime is unknown is F_{17}. The largest whose status has been settled is F_{1945}, and it is composite. Above $t = 4$, not a single value of t has thus far been found with F_t prime. Fermat's conjecture has met an unfortunate fate!

Let us return to the Mersenne numbers $M_p = 2^p - 1$. By good luck it happens that there is again a feasible test for primality. It is due to Lucas, as modified by Lehmer, and is as follows: Let $u_2 = 4$, $u_3 = 4^2 - 2 = 14$, $u_4 = 14^2 - 2 = 194$, and in general $u_{n+1} = u_n^2 - 2$. Then, M_p *is prime if and only if u_p is divisible by M_p.*

As in the case of Pepin's test for Fermat numbers, the proof that the Lucas-Lehmer test works is a little too difficult to present here. We illustrate it with $M_5 = 2^5 - 1 = 31$. (Of course, it is ridiculous to prove 31 to be prime this way. However the process it illustrates is not ridiculous for large Mersenne numbers.)

$$u_2 = 4, \quad u_3 = 14$$
$$u_4 = 194 \equiv 8 \bmod 31$$
$$u_5 \equiv 8^2 - 2 = 62 \equiv 0 \bmod 31$$

The outlook for actually using this test is quite the same as indicated in the discussion above for Fermat numbers. The method has in fact been employed to test all Mersenne numbers M_p with $p < 20{,}000$. On March 4, 1971, $M_{19{,}937}$ was found to be prime. (At the time of going to press, this stands as the largest known prime.) The computation was carried out by Bryant Tuckerman at IBM (on an IBM machine, of course!). The run for this one number took only 40 minutes, but of course many hours were spent checking all other primes up to 20,000. $M_{19{,}937}$ has 6,002 digits.

Twenty-three smaller Mersenne primes had been found previously. They are given by $p = 2, 3, 5, 7, 13, 17, 19, 31, 61, 89, 107, 127, 521, 607, 1279, 2203, 2281, 3217, 4253, 4423, 9689, 9941$, and 11213.

We shall now redeem the promise made early in this section to explain the connection between other topics in mathematics and Mersenne and Fermat primes. In a sentence: Mersenne primes are connected with perfect numbers, and Fermat primes are pertinent for the construction by ruler and compass of regular polygons.

Perfect numbers.   A number is called *perfect* if it is equal to the sum of its divisors. Here we count 1 as a divisor, but do not count the number itself. The early Greek mathematicians attached mystical properties to perfect numbers; because of this, curiosity about them continued through the ages to the present day.

The first perfect number is $6 = 1 + 2 + 3$. The next is $28 = 1 + 2 + 4 + 7 + 14$. The next two are 496 and 8128. Note that $6 = 2 \cdot 3$, $28 = 4 \cdot 7$, $496 = 16 \cdot 31$, $8128 = 64 \cdot 127$. A pattern is apparent, and was recognized long ago. Indeed the first half of the following theorem was proved by Euclid, the second half by Euler.

Theorem.   If q is a Mersenne prime, then $q(q + 1)/2$ is perfect. Any even perfect number is of this form.

The proof of this theorem is not difficult. We leave it as an informal exercise for the reader (or, of course, he can look it up in many books).

Concerning odd perfect numbers a great deal is known, indicating that if they exist they are enormous and have a very special form. (For instance, Pomerance announced in 1972 a proof that odd perfect numbers must have at least seven distinct prime factors. In 1973 Hagis and McDaniel showed that an odd perfect number must be larger than 10^{50} and must have a prime factor at least as large as 11213.) But all attempts to construct odd perfect numbers or to prove their nonexistence have thus far not been successful.

Regular polygons.   Constructing a regular polygon of n sides amounts to the same thing as constructing by ruler and compass the angle $360/n$ degrees $= 2\pi/n$ radians.

Early in Euclid's *Elements* a construction is given for bisecting any angle. It follows that if a regular polygon of n sides can be constructed, so can one of $2n$ sides. Thus the only real interest is the case where n is odd.

Of course the case $n = 3$ (an equilateral triangle) is trivial. Not so trivial is $n = 5$ (a regular pentagon), but it can be found in Euclid. There the matter rested for over 2000 years. In 1796, at the age of 18, Gauss found that a regular polygon of 17 sides is constructible. He later went on to give the exact result: *A regular polygon with n sides (n odd) is constructible by ruler and compass if and only if n is a product of distinct Fermat primes.*

Exercises

1. Let m and n be relatively prime positive integers. Prove that $2^m - 1$ and $2^n - 1$ are relatively prime.
2. (a) Let F_m and F_n be Fermat numbers with $m < n$. Prove that F_m divides $F_n - 2$.
   (b) Deduce that F_m and F_n are relatively prime.
3. (a) Show that the polynomial $x^r + 1$ is not divisible by $x - 1$ for any positive integer r.
   (b) Show that the polynomial $x^r + 1$ is not divisible by $x + 1$ for any odd positive integer r.
   (c) Show that the polynomial $x^r - 1$ is not divisible by $x + 1$ for any odd positive integer r.
4. Let q be a Fermat number. Show that $2^{q-1} \equiv 1 \bmod q$ holds whether or not q is a prime.
5. Let q be a Mersenne number. Show that $2^{q-1} \equiv 1 \bmod q$ holds whether or not q is a prime.
6. Prove that $a^{561} \equiv a \bmod 561$ holds for any integer a.
7. (a) Show that no number $2^n + 1$ with $n > 3$ is a power of 3.
   (b) Show that no number $2^n - 1$ with $n > 2$ is a power of 3.

(c) Suppose that $(3^m - 1)/2^r$ is prime, where 2^r is the highest power of 2 dividing $3^m - 1$. Prove that m must be a prime.

8. Use the identity

$$4x^4 + 1 = (2x^2 - 2x + 1)(2x^2 + 2x + 1)$$

to factor $(2^{58} + 1)/5$. (This number was first factored by Landry in 1869.)

9. Let $q = 2^r + 1$. Prove that $2^{2r} \equiv 1 \bmod q$ and that $2r$ is the smallest integer with this property.

10. What is the next power of 2 after 64 to have the property that both of its next neighbors are composite?

Chapter 3

Permutations

If we look back at what has gone on up to this point we see that at the outset the material was very abstract, dealing with rather general notions in the broad setting of set theory, and then became somewhat more concrete—with a discussion of the integers, prime numbers, and particular properties of particular types of integers. We now hit a road somewhere between the abstract and the concrete. This chapter is devoted to permutations, that is, one-to-one correspondences of a finite set onto itself.

Although our principal concern shall be with finite sets, in the initial material the finiteness of the set plays no role and the assumption of finiteness gives us no gain. Consequently we start out working in any set.

A remark or two about notation: In Chapter 1, where functions were introduced, given a function f from A to B we denoted this by $f: A \rightarrow B$ and wrote $f(a)$ for the value of f at the element a of A. We shall continue doing so in what follows. We shall also interchangeably use the terms *function, map,* or *mapping*.

In almost everything to be done we shall be interested in mappings of a set S into or onto itself. For ease of notation we shall consistently use lower-case Latin letters for elements of S and lower-case Greek letters for mappings, except that we shall use i to denote the identity map.

Suppose then that $\sigma: S \to S$ and $\tau: S \to S$ are mappings of S into itself. We recall a notion introduced in Chapter 1, Section 3. Using the mappings σ and τ, we defined another mapping of S into S called the *product* or *composite* of σ and τ, which we denoted by $\sigma \circ \tau$, or simply $\sigma\tau$. It was defined by specifying its action on every element $a \, \varepsilon \, S$ by the rule $(\sigma\tau)(a) = \sigma(\tau(a))$.

For the present purposes the product of two mappings of S into itself will be of the utmost importance. However, we make a change in the definition (the reason for this is indicated below) and, in fact, the change also affects the notation previously used. Given mappings σ, τ of S into itself we define $\sigma \circ \tau: S \to S$ by means of $(\sigma \circ \tau)(a) = \tau(\sigma(a))$ for every $a \, \varepsilon \, S$. Note that what we have done is to change the order of doing things, from that which we did earlier. Here, in $\sigma \circ \tau$ we first apply σ and then τ in contradistinction to what was done in Chapter 1, where it meant first apply τ and then σ. Note, too, that although we initially use the symbol $\circ$ for the sake of clarity in the early part of this chapter, we shall later drop it and write $\sigma \circ \tau$ as $\sigma\tau$.

What is the reason for introducing this change in the definition? Basically it boils down to the fact that we want to multiply permutations (one-to-one mappings of a set onto itself) from left to right—as is almost universally done by algebraists—rather than from right to left.

To avoid any misunderstanding, we repeat: In defining $\sigma \circ \tau$ in our present context, σ and τ being mappings of S into itself, we do so via $(\sigma \circ \tau)(a) = \tau(\sigma(a))$ for every $a \, \varepsilon \, S$; that is, first act with σ and then with τ.

Let us see this in a specific example. Let S be the set of integers and let $\sigma: S \to S$ be defined by $\sigma(n) = n + 3$ for any $n \, \varepsilon \, S$; let $\tau: S \to S$ be defined by $\tau(m) = m^2$ for any $m \, \varepsilon \, S$. What is $(\sigma \circ \tau)(14)$? According to our definition,

$$(\sigma \circ \tau)(14) = \tau(\sigma(14)) = \tau(14 + 3) = \tau(17) = 17^2 = 289$$

On the other hand,

$$(\tau \circ \sigma)(14) = \sigma(\tau(14)) = \sigma(14^2) = \sigma(196) = 196 + 3 = 199$$

Note that $\sigma \circ \tau \neq \tau \circ \sigma$, a phenomenon that we noticed several times earlier in Chapter 1, Section 3.

The object that we shall study is the set of mappings of a set S into itself; more precisely, we shall concern ourselves with the set of all one-to-one mappings of S *onto* itself. Our objective will be to study the behavior of this latter set with respect to the product or composite of functions discussed above.

1. The Set *A(S)*

Let S be a nonempty set. By $A(S)$ we shall mean the set of all one-to-one mappings of S onto itself. In Section 3 of Chapter 1 we discussed $A(S)$ and proved some properties for it. We want to transfer what we did there into our present setting. What has changed? Only the definition of the product, and even there the change is only in the order of doing things, which does not really affect anything we are about to say. Let us recall some of the material developed in Section 3 of Chapter 1. There we showed that the product of two one-to-one mappings of a set S onto itself is again a one-to-one mapping of S onto itself. In the present context this translates into: $\sigma, \tau \varepsilon A(S)$ implies $\sigma \circ \tau \varepsilon A(S)$. We describe this by saying that $A(S)$ is *closed* with respect to our product. Furthermore, we showed that the *associative law*, $\sigma \circ (\tau \circ \mu) = (\sigma \circ \tau) \circ \mu$, holds for any three elements in $A(S)$. The identity map on S, which we wrote as i or i_s, was shown to be in $A(S)$ and to satisfy the basic property $i \circ \sigma = \sigma \circ i = \sigma$ for every $\sigma \varepsilon A(S)$. Finally, we showed that if σ is a one-to-one mapping of S onto itself; that is, if $\sigma \varepsilon A(S)$, then there exists in $A(S)$ an element, written as σ^{-1}, such that $\sigma \circ \sigma^{-1} = \sigma^{-1} \circ \sigma = i$. Thus elements of $A(S)$ have inverses in $A(S)$.

We summarize the last paragraphs into the following theorem.

Theorem 3.1. $A(S)$ enjoys the following properties:

1. $\sigma, \tau \varepsilon A(S)$ implies $\sigma \circ \tau \varepsilon A(S)$.
2. $\sigma, \tau, \mu \varepsilon A(S)$ implies $\sigma \circ (\tau \circ \mu) = (\sigma \circ \tau) \circ \mu$.
3. There exists an element $i \varepsilon A(S)$ such that $\sigma \circ i = i \circ \sigma = \sigma$ for every $\sigma \varepsilon A(S)$.
4. Given $\sigma \varepsilon A(S)$, there exists an element $\sigma^{-1} \varepsilon A(S)$ such that $\sigma \circ \sigma^{-1} = \sigma^{-1} \circ \sigma = i$.

The four properties listed in the statement of Theorem 3.1 are those that define the abstract notion of a *group*. The special case of a group where the operation is commutative was touched on in Sections 2 and 3 of Chapter 2, where we were concerned with the additive properties of the integers. We shall investigate the general concept of a group in Chapter 4.

In the present chapter we shall content ourselves with a study of the particular group $A(S)$. This in fact follows the historical course of events. The group $A(S)$ was studied quite intensively in the late eighteenth century and in the early nineteenth century. Considerably later, in the middle part of the nineteenth century, the general notion of a group was introduced.

Exercises

1. Show that if $\sigma_0 \varepsilon A(S)$ commutes with all $\sigma \varepsilon A(S)$ (i.e., $\sigma_0 \circ \sigma = \sigma \circ \sigma_0$) then $\sigma_0 = i$, provided S has more than two elements. (See Exercises 34 and 35, p. 32.)

2. If $a_1 \ \varepsilon \ S$, let M be the set of all σ in $A(S)$ with $\sigma(a_1) = a_1$. Prove:
   (a) $\sigma, \tau \ \varepsilon \ M$ implies that $\sigma \circ \tau \ \varepsilon \ M$.
   (b) If $\sigma \ \varepsilon \ M$, then $\sigma^{-1} \ \varepsilon \ M$.

3. Show that if $a_1 \neq a_2$ are two distinct elements of S then there is a $\tau \ \varepsilon \ A(S)$ such that $\tau(a_1) = a_2$.

4. Let S be an *infinite* set and let F be the set of all $\sigma \ \varepsilon \ A(S)$ such that $\sigma(x) \neq x$ for only a *finite* number of x in S. Prove:
   (a) $\sigma, \tau \ \varepsilon \ F$ implies $\sigma \circ \tau \ \varepsilon \ F$.
   (b) $\sigma \ \varepsilon \ F$ implies $\sigma^{-1} \ \varepsilon \ F$.

5. Let S consist of the elements x_1, x_2, x_3, and let $\sigma: S \to S, \tau: S \to S$ be defined by $\sigma(x_1) = x_2, \sigma(x_2) = x_3, \sigma(x_3) = x_1$ and $\tau(x_1) = x_3$, $\tau(x_2) = x_2, \tau(x_3) = x_1$. Prove:
   (a) $\sigma \circ \tau \neq \tau \circ \sigma$
   (b) $(\sigma \circ \sigma) \circ \sigma = i$
   (c) $\tau \circ \tau = i$
   (d) $\tau \circ \sigma = \sigma^{-1} \circ \tau$

2. Cycle Decomposition

The specific nature of the set S has not entered into our discussion of $A(S)$ nor has it made its presence felt in any way. Now, for the first time, we shall restrict S.

Henceforth S shall be a *finite, nonempty* set having the n distinct elements $x_1, x_2, \ldots, x_n$. In this case $A(S)$ is usually written as S_n and is called the *symmetric group of degree n*. The elements of S_n are called *permutations* or, when one wants to be especially precise, permutations of degree n. Hence a permutation of degree n is nothing more or less than a one-to-one mapping of the set $S = \{x_1, \ldots, x_n\}$ onto itself, or in other words, a shuffle or rearrangement of these elements. As was indicated in Chapter 1, Section 3, S_n has $n!$ elements.

We shall often be faced with the need to discuss some particular element of S_n. In order to do this we shall need a device or notational scheme for representing this element. The device we shall use is that of being utterly explicit. We represent a permutation by writing, in some fashion, what it does to every element of S. Before describing it in general we illustrate what shall be done with an example.

Let $\sigma \ \varepsilon \ S_4$ be that permutation with:

$\sigma: x_1 \to x_2$
$\quad \ x_2 \to x_3$
$\quad \ x_3 \to x_1$
$\quad \ x_4 \to x_4$

We write this schematically as

$$\sigma = \begin{pmatrix} x_1 & x_2 & x_3 & x_4 \\ x_2 & x_3 & x_1 & x_4 \end{pmatrix}$$

Note that in the first row we have listed the four elements of S and that below each element, in the second row, we have written its image under σ. Since what counts is what is under each element in the first row, we could equally have written

$$\sigma = \begin{pmatrix} x_3 & x_1 & x_4 & x_2 \\ x_1 & x_2 & x_4 & x_3 \end{pmatrix}$$

Instead of discussing S_4 let us now move on to the general situation, namely S_n. Let σ be a permutation in S_n. Given $x_k \in S$, σ sends x_k again into S, and thus σ sends x_k to x_{i_k}, say. Because σ is one-to-one, if $x_k \neq x_m$, then $\sigma(x_k) \neq \sigma(x_m)$; that is, $x_{i_k} \neq x_{i_m}$. We represent σ symbolically as

$$\sigma = \begin{pmatrix} x_1 & x_2 & \cdots & x_n \\ x_{i_1} & x_{i_2} & \cdots & x_{i_n} \end{pmatrix}$$

Just as in the special case of the element in S_4 treated above, in the second row under the element x_k we put its image under σ, namely x_{i_k}.

A few remarks are in order. To begin with, since σ is one-to-one and onto, in the second row we find each of $x_1, \ldots, x_n$ once and only once (in scrambled order). Next, the order in which we write the elements in the first row is not important; what is important is only what is under each element, written in the second row. Thus we could equally well represent σ as

$$\sigma = \begin{pmatrix} x_2 & x_3 & x_1 & x_4 & \cdots & x_n \\ x_{i_2} & x_{i_3} & x_{i_1} & x_{i_4} & \cdots & x_{i_n} \end{pmatrix}$$

for instance.

Whatever else we have done, we now have a way of writing out each permutation in S_n in a somewhat compact form. Is it, however, the best way? A moment's perusal shows that there is a certain wastage in our symbols. What purpose is there in writing the x's? Why not drop them and merely represent x_k by its subscript k? Doing this, we would write the permutation

$$\begin{pmatrix} x_1 & x_2 & x_3 & x_4 \\ x_2 & x_3 & x_1 & x_4 \end{pmatrix}$$

in S_4 as

$$\begin{pmatrix} 1 & 2 & 3 & 4 \\ 2 & 3 & 1 & 4 \end{pmatrix}$$

For the general

$$\begin{pmatrix} x_1 & x_2 & \cdots & x_n \\ x_{i_1} & x_{i_2} & \cdots & x_{i_n} \end{pmatrix}$$

in S_n we would now write

$$\begin{pmatrix} 1 & 2 & \cdots & n \\ i_1 & i_2 & \cdots & i_n \end{pmatrix}$$

With this notation what is the identity element i of S_n? It is merely

$$\begin{pmatrix} 1 & 2 & \cdots & n \\ 1 & 2 & \cdots & n \end{pmatrix}$$

How do we calculate the product of two permutations of S_n in this nota-tion? It will be a little awkward to keep using the $\circ$ for the product so *henceforth we shall drop it* and write $\sigma \circ \tau$ merely as $\sigma\tau$. To calculate $\sigma\tau$, since this means first apply σ and to the result apply τ, we just follow through each element. We clarify this. Let

$$\sigma = \begin{pmatrix} 1 & 2 & 3 & 4 \\ 2 & 3 & 1 & 4 \end{pmatrix}, \quad \tau = \begin{pmatrix} 1 & 2 & 3 & 4 \\ 4 & 3 & 2 & 1 \end{pmatrix}$$

be in S_4. We see what

$$\sigma\tau = \begin{pmatrix} 1 & 2 & 3 & 4 \\ 2 & 3 & 1 & 4 \end{pmatrix} \begin{pmatrix} 1 & 2 & 3 & 4 \\ 4 & 3 & 2 & 1 \end{pmatrix}$$

does to each of 1, 2, 3, and 4 (i.e., x_1, x_2, x_3, and x_4). Under σ, 1 goes into 2 and under τ, 2 goes into 3; hence $\sigma\tau$ sends 1 into 3. Similarly $\sigma\tau$ sends 2 into 2, 3 into 4, and 4 into 1. Thus

$$\begin{pmatrix} 1 & 2 & 3 & 4 \\ 2 & 3 & 1 & 4 \end{pmatrix} \begin{pmatrix} 1 & 2 & 3 & 4 \\ 4 & 3 & 2 & 1 \end{pmatrix} = \begin{pmatrix} 1 & 2 & 3 & 4 \\ 3 & 2 & 4 & 1 \end{pmatrix}$$

which is easily seen visually by threading through each symbol, 1, 2, 3, 4. And this is of course the story in S_n.

Knowing the symbol for σ what is that of σ^{-1}? Easy, flip the second row of σ to become the first and the first to become the second (prove!). An example both explains and reveals this. Let

$$\sigma = \begin{pmatrix} 1 & 2 & 3 & 4 \\ 2 & 3 & 1 & 4 \end{pmatrix}$$

we claim that

$$\sigma^{-1} = \begin{pmatrix} 2 & 3 & 1 & 4 \\ 1 & 2 & 3 & 4 \end{pmatrix} = \begin{pmatrix} 1 & 2 & 3 & 4 \\ 3 & 1 & 2 & 4 \end{pmatrix}$$

Is it? A simple check shows

$$\begin{pmatrix} 1 & 2 & 3 & 4 \\ 2 & 3 & 1 & 4 \end{pmatrix} \begin{pmatrix} 1 & 2 & 3 & 4 \\ 3 & 1 & 2 & 4 \end{pmatrix} = \begin{pmatrix} 1 & 2 & 3 & 4 \\ 1 & 2 & 3 & 4 \end{pmatrix} = i$$

as desired.

In stripping our notational device of all the x's we have carried out one economy in writing. However, we still have wastage. The first row can always be taken to be 123 . . . *n*. Why not get rid of one row? Better still, we shall produce an even more condensed way of writing permutations. This we do now. We first do it for the special case of our old friend

$$\sigma = \begin{pmatrix} 1\ 2\ 3\ 4 \\ 2\ 3\ 1\ 4 \end{pmatrix}$$

in S_4. Pick some element in S, say 1; what does 1 go into under σ? We see it is 2. We write down (12 —. Now, what does 2 go into under σ? We see it is 3; we write (123 —. What does 3 go into under σ? We see it is 1; that is, we have returned to our starting point. We indicate this by closing the bracket and writing (123). What element is not accounted for? The element 4. Under σ, 4 goes into 4; we write this as (4). We now write

$$\sigma = \begin{pmatrix} 1\ 2\ 3\ 4 \\ 2\ 3\ 1\ 4 \end{pmatrix} = (123)(4)$$

and interpret the symbol as: In any given parenthesis an element goes into the one on its right under σ, except for the element preceding the parenthesis ')' which swings around. For instance, in S_8 the permutation (1476)(583)(2) would be the permutation

$$\begin{pmatrix} 1\ 2\ 3\ 4\ 5\ 6\ 7\ 8 \\ 4\ 2\ 5\ 7\ 8\ 1\ 6\ 3 \end{pmatrix}$$

which sends: $1 \to 4$, $4 \to 7$, $7 \to 6$, $6 \to 1$, $5 \to 8$, $8 \to 3$, $3 \to 5$, $2 \to 2$.

Let us examine more closely the procedure just described. Let S be a finite set (say with n elements) and let σ be a permutation of S. Take an arbitrary element of S, say a_1. Apply σ to a_1, getting an element we shall call a_2; then apply σ to a_2, getting a_3, etc. In other words, repeated action of σ on a_1 gives rise to a sequence $a_1, a_2, a_3, \ldots$ of elements of S. Now, sooner or later, there has to be a repetition in this sequence. In fact, as soon as we reach the element a_{n+1} a repetition must have occurred. (We can quote the Pigeon-Hole Principle to argue that $n + 1$ objects are being placed in n holes—the n elements of S—and two must fall in the same hole.) Now the fact that σ is one-to-one places a certain restriction on the manner in which the first repetition occurs. Let us illustrate. Suppose $a_1 = 4$, $a_2 = \sigma(4) = 2$, $a_3 = \sigma(2) = 9$, and so on, giving us the distinct elements

4, 2, 9, 5, 7, 6

and suppose a repetition pops up for the first time when we reach $\sigma(6)$. Is it possible, for instance, that $\sigma(6) = 9$? This would make the sequence now look like

4, 2, 9, 5, 7, 6, 9

But observe that $\sigma(2)$ and $\sigma(6)$ would then both be equal to 9, whereas σ

was supposed to be one-to-one. It is likewise impossible for $\sigma(6)$ to be 2, 5, 7, or 6. Thus the only way for a repetition to occur at this point is for $\sigma(9)$ to reach back to the original element 4. The sequence now reads

$$4, 2, 9, 5, 7, 6, 4$$

There is no point in continuing the sequence further; the numbers 4, 2, 9, 5, 7, 6 will just repeat. The effect of σ on these elements is now indicated by the notation we introduced above: (429576). We repeat that this symbol signifies that each of 4, 2, 9, 5, and 7 goes into its right neighbor, and 6 circles back to 4. We call (429576) a *cycle,* and because there are 6 elements being circulated in this particular cycle, we call it a 6-cycle.

In the general case, where we start with a_1 and let σ act on it repeatedly, we will get distinct elements for a while, say $a_1, a_2, \ldots, a_m$, and then when a repetition occurs it has to be via the equation $\sigma(a_m) = a_1$. This is argued precisely as above. The cycle has now closed as $(a_1 a_2 \cdots a_m)$, an *m*-cycle. Note that for each i from 1 to $m - 1$ we have $\sigma(a_i) = a_{i+1}$, and then $\sigma(a_m) = a_1$.

It is possible that all the n elements of S have shown up in the cycle $(a_1 a_2 \cdots a_m)$. In that case the action of σ on S is perfectly clear, and our task is done. If, on the contrary, there are elements that have not yet appeared, we do the obvious by starting the operation again with a fresh element b_1. This will result in building up a new cycle, say $(b_1 b_2 \cdots b_r)$. The length r of this cycle has nothing to do with the length m of the previous cycle; these can be absolutely any positive integers. The following point is now very important: None of the b's can be equal to any of the a's. This again follows from the fact that σ is one-to-one, as the reader should verify. What we have achieved so far is written as the product of the two disjoint cycles:

$$(a_1 a_2 \cdots a_m)(b_1 b_2 \cdots b_r)$$

and we have thus accounted for $m + r$ of the n elements of S. It is now clear how the procedure is to be continued to the bitter end. The result is the following theorem.

Theorem 3.2.   Every permutation in S_n can be written as a product of disjoint cycles.

Let us work out one more example. Take the permutation given in the first notation by

$$\begin{pmatrix} 1\ 2\ 3\ 4\ 5\ 6\ 7\ 8\ 9 \\ 4\ 5\ 3\ 6\ 9\ 7\ 1\ 8\ 2 \end{pmatrix}$$

Starting with 1 we trace the sequence of images 4, 6, 7 and close the cycle (1467). Then 2 leads to the cycle (259). The remaining elements 3 and 8 are left fixed, yielding the short cycles (3) and (8). Thus the permutation, as a product of disjoint cycles, is given by

(1467)(259)(3)(8)

We introduce one further abbreviation. We really need not write (3) and (8), if it is agreed that omitted letters are fixed under the permutation; they are called "fixed points." Our final answer is thus (1467)(259).

This procedure of omitting one-element cycles (i.e., fixed points, or elements fixed under σ) cannot quite be pushed to a *reductio ad absurdum*. If the permutation is the identity i, we have to write something, and the customary agreement is to write $i = (1)$.

REMARK. The breakup of a permutation into disjoint cycles can also be accomplished by using equivalence relations. For the crucial step in doing this, see Exercise 6.

If σ, τ are two cycles that are disjoint, then it is easy to see that $\sigma\tau = \tau\sigma$; thus the order in which the various cycles appear in the decomposition given in Theorem 3.2 does not matter. Note also that, for instance, the 4-cycle (1234) is the same as (3412)—in fact any cycle does not change if we permute its elements cyclically.

Consider the 3-cycle (123); as is immediately verified, (123) = (12) (13). In fact, given any m-cycle $(i_1 i_2 \cdots i_m)$ with $m \geqq 2$ and $i_1, \ldots, i_m$ distinct, then $(i_1 i_2 \cdots i_m) = (i_1 i_2)(i_1 i_3) \cdots (i_1 i_m)$. Thus for any m, an m-cycle is always a product of 2-cycles. Therefore 2-cycles play a special role and for this reason they are given a name of their own; they are called *transpositions*. Any cycle has just been shown to be a product of transpositions. Since, by Theorem 3.2, every permutation is a product of disjoint cycles, each of which is a product of transpositions, we see that every permutation is a product of transpositions. We have proved the important result

Theorem 3.3. Every permutation is a product of transpositions.

What the theorem says is eminently reasonable: To carry out any rearrangement of $x_1, \ldots, x_n$ we can do it in a series of interchanges of two elements at a time.

Note that there is no uniqueness, either in the transpositions or in their number, in the decomposition of Theorem 3.3. For instance (1234) = (12)(13)(14) = (23)(24)(13)(23)(13). However, as we shall see in the next section, the evenness or oddness of the number of transpositions needed to decompose a given permutation is an invariant of that permutation.

We are now going to examine what happens when we iterate a permutation σ. For this purpose it will be handy to use the short notation σ^m for the result of applying σ m times. Explicitly, we define σ^2 to be $\sigma\sigma$, σ^3 to be $\sigma\sigma\sigma$ (by the associative law it makes no difference how we group the three σ's), etc. We shall take for granted that the "usual" rules for manipulating powers are valid. But there is one rule that is especially important, and we single it out for attention: It says that if σ and τ commute, then $(\sigma\tau)^m = \sigma^m\tau^m$. Note that this equation usually will not hold if σ and τ do not commute. For instance, if we assume it for $m = 2$ we get $\sigma\tau\sigma\tau = \sigma^2\tau^2$; we can cancel σ on the left and τ on the right to get $\tau\sigma = \sigma\tau$. So the relation $(\sigma\tau)^m = \sigma^m\tau^m$ with $m = 2$ actually forces commutativity.

More generally, if $\sigma_1, \ldots, \sigma_k$ are permutations, any two of which commute, (i.e., $\sigma_t\sigma_j = \sigma_j\sigma_t$ for all t and j) then $(\sigma_1 \cdots \sigma_k)^m = \sigma_1{}^m \cdots \sigma_k{}^m$.

Now let us consider a transposition σ, for example (46) in S_7, and work out σ^2. Schematically we have, where every arrow denotes an application of σ,

$$
\begin{aligned}
1 &\to 1 \to 1 \\
2 &\to 2 \to 2 \\
3 &\to 3 \to 3 \\
4 &\to 6 \to 4 \\
5 &\to 5 \to 5 \\
6 &\to 4 \to 6 \\
7 &\to 7 \to 7
\end{aligned}
$$

Hence σ^2 sends every element to itself; that is, $\sigma^2 = i$. This is plainly true for any transposition.

A similar computation, which we leave to the reader, shows that if σ is an m-cycle, then $\sigma^m = i$ and $\sigma^k \neq i$ for $1 \le k < m$.

Now let τ be any permutation in S_n. By Theorem 3.2, $\tau = \sigma_1\sigma_2 \cdots \sigma_k$ where the σ's are disjoint cycles. Because they are disjoint, they commute. Suppose that σ_j is an m_j-cycle ($j = 1, 2, \ldots, k$). We then have $\sigma_j{}^{m_j} = 1$. Write $m = m_1 m_2 \cdots m_k$. Then

$$\sigma_j{}^m = \sigma_j{}^{m_1 m_2 \cdots m_k} = (\sigma_j{}^{m_j})^{m_1 m_2 \cdots m_{j-1} m_{j+1} \cdots m_k} = i^{m_1 m_2 \cdots m_{j-1} m_{j+1} \cdots m_k} = i$$

But then $\tau^m = \sigma_1{}^m\sigma_2{}^m \cdots \sigma_k{}^m = ii \cdots i = i$. In fact we need not necessarily use m, for any integer divisible by each of $m_1, m_2, \ldots, m_k$ will do. Hence the least common multiple of $m_1, \ldots, m_k$—call it $\bar{m}$ for the moment—gives us $\tau^{\bar{m}} = i$. *We leave as Exercise 5 the proof that no positive integer smaller than $\bar{m}$ will have this property.*

Let us call the *order* of τ the least positive integer r such that $\tau^r = i$. By the above, $r = \bar{m}$, the least common multiple of the $m_1, \ldots, m_k$, the lengths of the various cycles of τ.

We compute the orders of some permutations. Consider the permutation

$$
\begin{pmatrix}
1\,2\,3\,4\,5\,6\,7\,8\,9 \\
3\,9\,5\,1\,4\,7\,8\,6\,2
\end{pmatrix}
$$

First we break it up as (1354)(678)(29). The lengths of its disjoint cycles are 4, 3, and 2 and since the least common multiple of these is 12 we see that the order of the permutation is 12.

What is the order of (12)(24)(431)? As it stands, it is not the product of disjoint cycles, so we first accomplish this. Now (12)(24)(431) = (13)(24) and we see that its order is 2.

Suppose we have a deck of cards with eight cards numbered 1 to 8 from top to bottom. We have a machine that carries out the following shuffle of the cards: It cuts the deck in half, puts the bottom half on top of the top half, and then takes the resulting top card and puts it at the bottom. This is the shuffle. How many times must this shuffle be repeated to get the cards back in their original order 1, 2, . . . , 8?

The cards are originally in the order 1, 2, 3, 4, 5, 6, 7, 8. After the cut we have 5, 6, 7, 8, 1, 2, 3, 4, and putting the top card last we end up with 6, 7, 8, 1, 2, 3, 4, 5. Thus the card numbered 6, which was originally in the 6th slot, now occupies the 1st place; the card numbered 7, which was originally in the 7th slot, now occupies the 2nd place, etc. So the shuffle effects the permutation $6 \to 1, 7 \to 2, 8 \to 3, 1 \to 4, 2 \to 5, 3 \to 6, 4 \to 7, 5 \to 8$, that is

$$\begin{pmatrix} 1\ 2\ 3\ 4\ 5\ 6\ 7\ 8 \\ 4\ 5\ 6\ 7\ 8\ 1\ 2\ 3 \end{pmatrix}$$

Decomposing this permutation as a product of disjoint cycles we get (14725836), an 8-cycle which is of order 8. Hence repeating the shuffle 8 times brings us back to the original order.

Show that for a deck of 10 cards using the shuffle described above gives a shuffle that must be repeated five times to bring the cards back to their original order. See Exercise 19 and Section 4 for other shuffles.

Exercises

1. If σ, τ are disjoint cycles, prove that $\sigma\tau = \tau\sigma$.
2. If σ, τ are permutations whose respective cycle decompositions have no element in common, prove that $\sigma\tau = \tau\sigma$.
3. What is σ^{-1} if σ is the m-cycle $(i_1 i_2 \cdots i_m)$?
4. If σ is an m-cycle, prove that $\sigma^m = i$ and $\sigma^k \neq i$ for $1 \leq k < m$.
5. Show that the order of a permutation is the least common multiple of the lengths of its cycles.
6. Let S be a finite set and let $\sigma \varepsilon A(S)$. For $a, b \varepsilon S$ define $a \sim b$ if $b = \sigma^k(a)$ for some integer $k \geq 0$. Prove that this is an equivalence relation on S.
7. Find the decomposition into disjoint cycles of:
   (a) $\begin{pmatrix} 1\ 2\ 3\ 4\ 5\ 6 \\ 3\ 2\ 5\ 6\ 1\ 4 \end{pmatrix}$
   (b) (12)(1342)(435)(25)
   (c) $(123)^{-1}(14536)(123)$

8. In Exercise 7 determine the orders of the permutations in parts (a), (b), and (c).

9. Decompose the permutations in Exercise 7 into products of transpositions.

10. If $\sigma \neq \tau$ are two transpositions, prove:
    (a) $\sigma\tau$ is of order 2 if σ and τ have no letter in common.
    (b) $\sigma\tau$ is a 3-cycle (and so, of order 3) if σ and τ have exactly one letter in common.

11. Let σ be a transposition and τ a 3-cycle.
    (a) If σ and τ overlap in exactly one letter, prove that $\sigma\tau$ is a 4-cycle.
    (b) If σ and τ overlap in two letters, prove that $\sigma\tau$ is a transposition.

12. If σ is an m-cycle, prove that $\tau^{-1}\sigma\tau$ is an m-cycle for any permutation τ.

13. Two permutations are said to have the *same cycle structure* if in their decomposition as products of disjoint cycles the length (and the number of times a given length appears) of their cycles is the same. (For instance, (12)(34)(567) and (35)(17)(426) have the same cycle structure whereas (12)(34)(567) and (13)(245) do not.) If σ and τ are any permutations in S_n, prove that σ and $\tau^{-1}\sigma\tau$ have the same cycle structure.

14. Prove that σ and σ^{-1} always have the same cycle structure.

15. If σ, τ are two m-cycles, prove that we can find a permutation λ such that $\tau = \lambda^{-1}\sigma\lambda$.

16. If σ, τ have the same cycle structure, prove that there is a permutation λ such that $\tau = \lambda^{-1}\sigma\lambda$.

17. Let $\sigma = (123456)$. Decompose each of σ^2, σ^3, σ^4, σ^5 into a product of disjoint cycles.

18. Let $\sigma_0 \ \varepsilon \ A(S)$. Define the mapping $f: A(S) \to A(S)$ by $f(\tau) = \sigma_0^{-1}\tau\sigma_0$ for $\tau \ \varepsilon \ A(S)$. Prove:
    (a) f maps $A(S)$ *onto* itself and is one-to-one.
    (b) $f(\tau\gamma) = f(\tau)f(\gamma)$ for all $\tau, \gamma \ \varepsilon \ A(S)$.

19. Let a shuffle of a deck of 10 cards be made as follows: The top card is put at the bottom, the deck is cut in half, the bottom half is placed on top of the top half, and then the resulting bottom card is put on top. How many times must this shuffle be repeated to get the cards back in the initial order?

3. Even and Odd

In the preceding section we saw that any permutation can be decomposed as a product of transpositions. However, this decomposition is not unique, as we saw in the example

$$(1234) = (12)(13)(14) = (23)(24)(13)(23)(13)$$

Nevertheless, we may wonder if there is anything about such decompositions that can be described, in any way, as unique.

Our first aim is to show that the *parity* of the number of transpositions used in writing a given permutation as a product of transpositions is indeed an invariant. What does this mean? It says that, given a certain permutation, if it has any decomposition as a product of an even number of transpositions, then any other decomposition whatever of this permutation must consist of an even number of transpositions; likewise, once a product of an odd number of transpositions, always a product of an odd number of transpositions. This will be proved in Theorem 3.5; however, we first need some preliminary results.

Permutations, in particular transpositions, need not commute. However some sort of substitute for commutativity does hold. For what is to follow, the next lemma is a simple but important step.

Lemma 3.1. If σ, τ are transpositions, then $\sigma\tau = \tau_1\sigma$ for *some* transposition τ_1.

Proof. Let $\sigma = (ij)$, $\tau = (ab)$ be the two transpositions. If the four letters i, j, a, and b are distinct, then σ and τ commute so $\sigma\tau = \tau\sigma$ and $\tau = \tau_1$ satisfies the conclusion of the lemma. Similarly, if $\sigma = \tau$, then $\sigma\tau = \tau\sigma$ so the result is correct. Finally, suppose that $\sigma \neq \tau$ and one of a or b is one of i or j, say $a = i$. Then $b \neq j$. We then see that $\sigma\tau = (ij)(ab) = (ij)(ib) = (jb)(ij) = \tau_1\sigma$, where $\tau_1 = (jb)$ is the transposition that does the trick. This proves the lemma.

What does the lemma say? It says that we can move any given transposition σ across any other one, from left to right, without changing σ, but possibly changing the other one into another transposition.

Theorem 3.4. The identity element of S_n *cannot* be written as a product of an *odd* number of transpositions.

Proof. Before going into the details of the proof let us give a rough description of what the proof shall be. The preceding lemma gives us a sort of commutativity for transpositions; we shall use it to pull all transpositions not involving a given particular letter to the right, all those involving that letter to the left. Once this is done we shall have a factorization of i, the identity of S_n, into a product of transpositions in a particular way. From this we will be able to finish the proof. We now proceed to actuate this rather vague description.

Suppose that we can write i as a product of an odd number of transpositions; then there is a shortest such representation (i.e., one involving as small an odd number of transpositions as possible). Let this minimal odd decomposition of i be $i = \tau_1\tau_2 \cdots \tau_{2k+1}$. Write $\tau_1 =$

(*ab*). Let τ_j be the first transposition not involving a. By Lemma 3.1 we can pull τ_j across τ_{j+1}, keeping τ_j the same but possibly changing τ_{j+1} into another transposition. Continue this process of pulling τ_j across all the transpositions to the right of it. This way we will have put τ_j at the right-hand end; in carrying out this "pulling across" we have not changed the number of transpositions involved.

Now, in the same way, take the first transposition not involving a and pull it to the second last position on the right. Continue the process; note that the number of transpositions used, namely $2k + 1$, stays the same. What have we accomplished? We now have i factored as $i = \sigma_1\sigma_2\cdots\sigma_m\sigma'_{m+1}\cdots\sigma'_{2k+1}$ where the transpositions σ_1, σ_2, ..., σ_m all involve a, whereas the transpositions $\sigma'_{m+1}, \ldots, \sigma'_{2k+1}$ do *not* involve a.

Now, since $\sigma_1, \ldots, \sigma_m$ all involve a, $\sigma_1 = (ab_1)$, $\sigma_2 = (ab_2)$, ..., $\sigma_m = (ab_m)$. If all the elements $b_1, b_2, \ldots, b_m$ are distinct, then, under $\sigma_1\sigma_2\cdots\sigma_m$, b_m goes into a; because $\sigma'_{m+1}, \ldots, \sigma'_{2k+1}$ do not involve a, they leave a fixed. Hence b_m goes into a under $\sigma_1\sigma_2\cdots\sigma_m\sigma'_{m+1}\cdots\sigma'_{2k+1}$. But this is nonsense, for we have supposed that $\sigma_1\sigma_2\cdots\sigma_m\sigma'_{m+1}\cdots\sigma'_{2k+1} = i$, and so carries b_m into b_m. In short, we cannot have this situation. Thus two of the elements $b_1, \ldots, b_m$ must be equal, say $b_r = b_s$ for some $1 \leq r < s \leq m$, and $\sigma_r = \sigma_s$.

As before, we can now pull σ_r across its right-hand neighbors, without changing the number of transpositions involved, to move it next to σ_s. But then σ_r and σ_s abut in the product, giving the term $\sigma_r\sigma_s = \sigma_s^2 = i$. This says that we have eliminated σ_r and σ_s from the product representation of i as an odd product of transpositions. In other words, we have shortened by two the representation of i as a product of an odd number of transpositions. This contradicts our initial assumption that we had the shortest odd product representing i. The proof is complete.

The hard work is over. Now we can easily get the basic result we are seeking, but first:

Definition. A permutation in S_n is called *odd* if it is a product of an odd number of transpositions and is called *even* if it is a product of an even number of transpositions.

We now prove:

Theorem 3.5. A permutation is either even or odd *but cannot be both.*

Proof. Given a permutation σ, we know it is a product of transpositions. Hence σ is a product of either an even number or an odd number of transpositions; that is, σ is either even or odd. All that remains is to show that it cannot be both.

Suppose that σ is both even and odd; then $\sigma = \tau_1 \cdots \tau_{2q} = \tau'_1 \cdots \tau'_{2r+1}$ where all the τ's, (τ')'s are transpositions. Since $(\tau'_{2r+1})^2 = i$, multiplying $\tau_1 \cdots \tau_{2q} = \tau'_1 \cdots \tau'_{2r+1}$ from the right by τ'_{2r+1} yields $\tau_1 \cdots \tau_{2q}\tau'_{2r+1} = \tau'_1 \cdots \tau'_{2r}$. Continue, to get $i = \tau_1 \cdots \tau_{2q} \tau'_{2r+1} \cdots \tau'_1$. But we know that i is not a product of an odd number of transpositions, so the theorem has been proved.

We leave the proof of the following theorem to the reader (Exercise 2).

Theorem 3.6.   The following are true:

1. A product of two even permutations is even.
2. A product of two odd permutations is even.
3. A product of an even and an odd permutation is odd.
4. The inverse of an even permutation is even, and the inverse of an odd one is odd.

Let us look at this theorem in a slightly different way. Given a permutation σ we define the *signature* of σ as follows: The signature of σ is 1 if σ is even, and the signature of σ is -1 if σ is odd. We write the signature of σ as $\Lambda(\sigma)$. The above theorem, in these terms, becomes

Theorem 3.7.   For $\sigma, \tau \; \varepsilon \; S_n$, $\Lambda(\sigma\tau) = \Lambda(\sigma)\Lambda(\tau)$.

We also leave the proof of this theorem to the reader (Exercise 2).

How many even and how many odd permutations are there in S_n? We claim that each set makes up one half of S_n; that is, there are $n!/2$ even ones and $n!/2$ odd ones (if $n > 1$). How do we see this? Let A_n be the set of even permutations and B_n the set of odd ones. We have a ready passage from A_n to B_n, namely, multiply every element in A_n by (12).

By part 3 of Theorem 3.6, multiplying any element of A_n by (12), which is odd, throws us into B_n. It is immediate that this is a one-to-one map of A_n into B_n. So A_n has at most as many elements as B_n. Now do the same thing to B_n; that is, map B_n into A_n by multiplying each of its elements by (12). By part 2 of Theorem 3.6 we do indeed map B_n into A_n this way. So, as above, we get that B_n has at most as many elements as A_n. Thus A_n and B_n have the same number of elements. We sum this up as

Theorem 3.8.   In S_n, $n > 1$ there are $n!/2$ even permutations and $n!/2$ odd permutations.

A_n is an interesting subset of S_n; it is called the *alternating group of degree n*. What properties does it enjoy? To begin with, if $\sigma, \tau \; \varepsilon \; A_n$, then

$\sigma\tau$ is in A_n as we have seen. Furthermore, if $\sigma \;\varepsilon\; A_n$, then $\sigma^{-1} \;\varepsilon\; A_n$ by part 4 of Theorem 3.6. So A_n contains the product of any two of its elements and the inverse of any element in it. The associative law holds for any three elements of A_n because it already holds in S_n. Finally $i \;\varepsilon\; A_n$. Summarizing we see that A_n satisfies all the properties in the conclusion of Theorem 3.1 describing the behavior of S_n. For this reason A_n is called a *subgroup* of S_n. We shall hear more about subgroups in the next chapter.

We close this section with a few words about a game played with a simple toy. This game seems to have been invented in the 1870s by the famous puzzle-maker Sam Loyd. It caught on and became the rage in the United States in the 1870s, and finally led to a discussion by W. Johnson in the scholarly journal, the *American Journal of Mathematics,* in 1879. It is often called the "fifteen puzzle." Our discussion will be without full proofs.

Consider a toy made up of 16 squares, numbered from 1 to 15 inclusive and with the lower right-hand corner blank.

1	2	3	4
5	6	7	8
9	10	11	12
13	14	15	

The toy is constructed so that squares can be slid vertically and horizontally, such moves being possible because of the presence of the blank square.

Start with the position shown and perform a sequence of slides in such a way that, at the end, the lower right-hand square is again blank. Call the new position "realizable." Question: What are all possible realizable positions?

What do we have here? After such a sequence of slides we have shuffled about the numbers from 1 to 15; that is, we have effected a permutation of the numbers from 1 to 15. To ask what positions are realizable is merely to ask what permutations can be carried out. In other words, in S_{15}, the symmetric group of degree 15, what elements can be reached via the toy? For instance, can we get

13	4	12	15
1	14	9	6
8	3	2	7
10	5	11	

At any rate one observation is easy to make. If one can carry out a given permutation it must be even. Consider it this way: Imagine that the blank square is numbered 16, for the moment, and that we are working in S_{16}. In our move the blank square (16) is moved around and comes back to the lower-right corner. As the blank (16) is moved horizontally or vertically, each one-square move is an interchange of 16 with some other number, hence is a transposition. We want to show that in reaching a realizable permutation there must have been an even number of such transpositions.

Imagine the board colored as a checker board, alternate squares in black and white. Each horizontal or vertical one-square move takes you from a square of one color to a square of the opposite color. Hence an odd number of one-square moves with a given piece will end on a square of opposite color, whereas an even number of such moves brings the piece back to a square of the same color. Now 16 started in the lower-right corner and ends up in the lower-right corner—hence no change of color—and so it was involved in an *even* number of one-square moves. Hence the permutation achieved is a product of an even number of transpositions in S_{16}, and so is an even permutation. Since it actually leaves 16 (the blank) fixed, it only permutes the first 15 numbers, and the permutation, considered as a permutation on 15 things, is therefore even (see Exercise 13).

Of greater difficulty is the fact that every even permutation is indeed realizable, but that is the case. No really easy proof seems to be known. Granted this, the number of realizable positions is 15!/2—a goodly number—the number of even permutations in S_{15}.

Returning to whether or not

13	4	12	15
1	14	9	6
8	3	2	7
10	5	11	

can be realized, we write it as a permutation

$$\begin{pmatrix} 1 & 2 & 3 & 4 & 5 & 6 & 7 & 8 & 9 & 10 & 11 & 12 & 13 & 14 & 15 \\ 5 & 11 & 10 & 2 & 14 & 8 & 12 & 9 & 7 & 13 & 15 & 3 & 1 & 6 & 4 \end{pmatrix}$$

Break this up as a product of disjoint cycles:

$$(1, 5, 14, 6, 8, 9, 7, 12, 3, 10, 13)(2, 11, 15, 4)$$

We have a product of an 11-cycle and a 4-cycle. However, an 11-cycle is even whereas a 4-cycle is odd (see Exercise 9); therefore their product is odd. Hence the given permutation is not realizable.

This gives us a general method for determining whether a given position is realizable. Write it as a permutation in S_{15}, decompose it into disjoint cycles, and then use Exercise 9 or Exercise 10 to determine the parity of the permutation. If the permutation is even, the position is realizable, if odd it is not. A collateral problem might be to find the most economical way of reaching an attainable position in the puzzle.*

Exercises

1. Determine the parity of the following permutations:
   (a) $\begin{pmatrix} 1\ 2\ 3\ 4\ 5\ 6\ 7\ 8 \\ 3\ 4\ 5\ 6\ 2\ 1\ 8\ 7 \end{pmatrix}$
   (b) $\begin{pmatrix} 1\ 2\ 3\ 4\ 5\ 6\ 7\ 8\ 9 \\ 3\ 4\ 5\ 6\ 2\ 1\ 8\ 7\ 9 \end{pmatrix}$
   (c) $(1274)(356)$
   (d) $(12)(13)(15)(346)(72)$
   (e) $(158)(25)(1367)(28)$
2. Prove Theorems 3.6 and 3.7.
3. If σ^m is odd and m is odd, prove that σ is odd.
4. Why is the identity permutation even?
5. For $n > 2$ show that any element in A_n can be written as a product of 3-cycles.
6. Exhibit two 5-cycles whose product is a 3-cycle.
7. If $\tau \ \varepsilon \ S_n$ is odd, show that $\sigma^{-1}\tau\sigma$ is odd for all $\sigma \ \varepsilon \ S_n$.
8. Let $W \subset S_n$ be such that $\sigma, \ \tau \ \varepsilon \ W$ implies $\sigma\tau \ \varepsilon \ W$. Prove that either all the elements of W are even permutations or else exactly half of the elements of W are even permutations.
9. Show that an m-cycle is an even permutation if m is odd and an odd permutation if m is even.
10. Show that a permutation is even if and only if its decomposition into disjoint cycles involves an even number of cycles of even length.
11. Prove that the following position in the 15 puzzle is not realizable:

10	4	5	7
8	6	11	9
14	13	12	15
1	3	2	

12. Show that the following position is realizable in the 15 puzzle:

*A recent reference for the 15 puzzle and related topics is Martin Gardner, *Sixth Book of Mathematical Games from Scientific American* (San Francisco: Freeman, 1971). Chapter 7 discusses other versions of the puzzle (for instance, where the blocks moved around need not be square). Item 9 of Chapter 20 discusses how to do it economically (with 8 instead of 15).

For a thorough history of the proof of Theorem 3.5 see T. L. Bartlow, "An Historical Note on the Parity of Permutations," *American Mathematical Monthly*, 79 (1972); 766–769. There is a large bibliography to which at least one more item can be added: S. Lang, *Basic Mathematics* (Reading, Mass.: Addison-Wesley, 1971), pp. 364–372.

12	15	13	14
11	1	8	10
6	7	2	9
5	3	4	

If you have a copy of the puzzle, carry it out explicitly.

13. Let $m > n$. If σ is a permutation in S_n, we can consider it as a permutation in S_m as follows: σ acts on $1, 2, \ldots, n$ as it did as an element of S_n and on $n + 1, n + 2, \ldots, m$, σ acts as the identity map; that is, $\sigma(j) = j$ for $n + 1 \le j \le m$.

Show that if $\sigma \ \varepsilon \ S_n$ is an odd permutation, then it is an odd permutation in all S_m for $m > n$; ditto if σ is even.

14. Let us formalize slightly what we did in Exercise 13. Let $m > n$. We define a mapping $f: S_n \to S_m$ as follows: If $\sigma \ \varepsilon \ S_n$, then $f(\sigma) = \tilde{\sigma} \ \varepsilon \ S_m$ is defined by $\tilde{\sigma}(j) = \sigma(j)$ for $j = 1, 2, \ldots, n$ and $\tilde{\sigma}(j) = j$ for $j > n$. Prove:
    (a) f is a one-to-one mapping of S_n into S_m.
    (b) f is not onto.
    (c) $f(\sigma\tau) = f(\sigma)f(\tau)$

15. Consider an 8×8 checker board. Prove that it is impossible to go from the upper-left corner to the lower-right corner—moving horizontally and vertically—going through each square once and *only* once.

16. In the following permutation the two missing letters on the bottom got lost, but it is remembered that the permutation is even. Restore the missing letters

$$\begin{pmatrix} 1 \ 2 \ 3 \ 4 \ 5 \ 6 \ 7 \\ 2 \ 7 \quad \ \ 6 \ 3 \ 4 \end{pmatrix}$$

17. Show that the product of two 5-cycles cannot be a 4-cycle.

18. The following is a known theorem: Any even permutation on $1, 2, \ldots, n$ can be written as a product of two n-cycles. Check it up to $n = 6$. Why do we insist that the permutation be even?

19. In a certain class there are 25 students, sitting at desks arranged in 5 rows of 5 each. One of the students suggests that each student change his or her seat, but only by moving forward, back, to the right, or to the left one seat. Can they do it?

20. The "15 puzzle" can be instructively cut down to a "5 puzzle," which still exhibits the key ideas. In this smaller puzzle the numbers 1, 2, 3, 4, 5 are inserted in the blank squares of

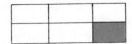

and the object is to reach

1	2	3
4	5	

Once again success just depends on whether the given permutation is even or odd. For

2	1	4
3	5	

check that the permutation is even and carry out the solution. (If you do this carefully, you should see a plan that can be extended to the 15 puzzle, or any size board.)

A brutal hand computation, not carefully checked, indicated that the most economical solutions require 18 moves. Can there be a solution with an odd number of moves?

4. The Interlacing Shuffle

There is a standard way of shuffling cards by dividing the deck into two halves and then interlacing the halves. It is used at card tables all over the world. Card sharps can perform it faultlessly with a deft flick of their wrists. Let us see what this shuffle looks like as a permutation.

We are given an even number of cards, numbered from 1 to $2n$. We divide them into two halves of n each: The first half carries the numbers 1 to n, and the second half those from $n + 1$ to $2n$. We shall interlace them as follows: Start with $n + 1$, the first card of the second pile; then take 1, the first card of the first pile; we continue alternating between the halves, the next few cards being $n + 2$, 2, $n + 3$, 3. The final two cards are $2n$, n. The $2n$ cards have now been rearranged and appear in the following order:

$$n + 1, 1, n + 2, 2, n + 3, 3, \ldots, 2n, n$$

For instance, for $n = 4$ ($2n = 8$), the new order for the eight cards is

$$5, 1, 6, 2, 7, 3, 8, 4$$

(Perhaps the reader feels that it would be more natural to interlace in the reverse order, so that for $n = 4$ we would get

$$1, 5, 2, 6, 3, 7, 4, 8$$

The result would be a closely related permutation for which it is also possible to give a concise description—see Exercise 4. We prefer our way of doing it simply because it works out a little more neatly.)

Let us set up the permutation that has occurred. We do this first for the special case of 8 cards ($n = 4$). The card numbered 5 has moved from the 5th position to the 1st position; the card numbered 1 has moved from the 1st position to the 2nd, etc. Thus the permutation, in the first of our two notations, is

$$\begin{pmatrix} 5\ 1\ 6\ 2\ 7\ 3\ 8\ 4 \\ 1\ 2\ 3\ 4\ 5\ 6\ 7\ 8 \end{pmatrix}$$

Let us restore the first row of the permutation to its natural order. We then get

$$\begin{pmatrix} 1\ 2\ 3\ 4\ 5\ 6\ 7\ 8 \\ 2\ 4\ 6\ 8\ 1\ 3\ 5\ 7 \end{pmatrix}$$

In this form, the pattern is clear. For the general case of $2n$ cards, the interlacing permutation (let us call it I_n) is given by

$$I_n = \begin{pmatrix} 1\ 2\ 3\ 4 \cdots n & n+1 & n+2 \cdots 2n \\ 2\ 4\ 6\ 8 \cdots 2n & 1 & 3 \cdots 2n-1 \end{pmatrix}$$

We can describe the effect of I_n quite explicitly: $I_n(i) = 2i$ for $1 \leq i \leq n$ and $I_n(i) = 2i - (2n+1)$ for $n + 1 \leq i \leq 2n$.

We can put this in a shorter, neater way. The subtraction of $2n + 1$ that occurs in the expression $2i - (2n + 1)$ suggests that we consider what is happening modulo $2n + 1$. Let us go back to $n = 4$ to look at this in detail. If we simply double the numbers 1, 2, 3, 4, 5, 6, 7, 8, we get 2, 4, 6, 8, 10, 12, 14, 16. Now in this case $2n + 1 = 9$. So if we reduce modulo 9, the last four numbers (10, 12, 14, and 16) will be cut down to 1, 3, 5, and 7, which is exactly what we want.

In just the same way the general case works, and we may summarize as follows: *The interlacing permutation I_n really acts on the $2n$ nonzero residues modulo $2n + 1$ (represented by 1, 2, . . . , $2n$) by doubling them.* This is certainly a pleasant connection between our previous study of number theory and our current study of permutations.

To better understand the interlacing permutation it is desirable to decompose it into a product of disjoint cycles. Unfortunately, our study of number theory was not carried quite far enough for us to do a complete job on this. We shall have to be content with some examples, remarks, and exercises. First we give a table up to $n = 8$.

n	I_{2n}, the interlacing permutation on 1, 2, . . . , $2n$
1	(1 2)
2	(1 2 4 3)
3	(1 2 4)(3 6 5)
4	(1 2 4 8 7 5)(3 6)
5	(1 2 4 8 5 10 9 7 3 6)
6	(1 2 4 8 3 6 12 11 9 5 10 7)
7	(1 2 4 8)(3 6 12 9)(5 1)(7 14 13 11)
8	(1 2 4 8 16 15 13 9)(3 6 12 7 14 11 5 10)

Here is one simple observation that is within our reach. Suppose that $2n + 1$ is a prime. By Fermat's little theorem we have $2^{2n} \equiv 1$ (mod $2n + 1$). This tells us that if the operation of doubling modulo $2n + 1$ is repeated $2n$ times we get back to the order in which we started. In other words, $I_{2n}{}^{2n}$ is the identity permutation. It follows that the order of I_{2n} divides $2n$, and that every cycle in I_{2n} has a size dividing $2n$. Note how this checks out in the table above. The values of n for which $2n + 1$ is a prime are given by $n = 1, 2, 3, 5, 6$, and 8. For $n = 1, 2, 5$, and 6, I_{2n} is a single $2n$-cycle. For $n = 3$ and 8 the $2n$ letters split into two n-cycles. As a matter of fact it can be seen (by arguments a little too advanced for this book) that when $2n + 1$ is a prime, I_{2n} splits into cycles of the same size. See Exercise 6 in this connection.

Exercises

1. Let T be the subset of the numbers $1, 2, \ldots, 2n$ consisting of those that are prime to $2n + 1$. Prove that I_{2n} carries the set T into itself (i.e., if $i \in T$, then $I_{2n}(i) \in T$). Verify this for $n = 4$ and 7 in the table above.
2. Find the next three values of n after $n = 8$ such that $2n + 1$ is prime, and work out the cycle structure of I_{2n} for them.
3. Show that I_{30} breaks up into six 5-cycles.
4. Consider the other interlacing shuffle mentioned in the text: the one that for $n = 4$ places the eight cards in the order

    1, 5, 2, 6, 3, 7, 4, 8

    Show that for a general value of n (i.e., there are $2n$ cards), this permutation is describable as follows: for $1 \leqq i \leqq n$, i is sent into $2i - 1$, and for $n + 1 \leqq i \leqq 2n$, i is sent into $2i - 2n$.
5. The late Ely Culbertson believed that probabilities in bridge are distorted by imperfect shuffling. On page 12 of his *Red Book on Play*, he says:

    Manufacturers of playing cards arrange them in suits of thirteen cards each. That alone is sufficient to increase tremendously the probability of dealing a thirteen card suit. Any player who will take the trouble to shuffle a new deck in such a manner that each card alternates with the other will find that after a few such shuffles he creates a pattern which results, *no matter how the cards are cut* (E. C.'s emphasis), in each player's receiving thirteen cards of a suit.

    Prove that exactly *four* successive interlacing shuffles will accomplish the objective.

    Try to generalize this result to an arbitrary deck of cards. (A general deck has r suits with s cards in each suit; for the usual deck, $r = 4$ and $s = 13$.)

6. Let p be an odd prime. We say that 2 is a *primitive root* of p if $2^h \equiv 1 \bmod p$ (which we know holds for $h = p - 1$ by Fermat's little theorem) does not hold for any $h < p - 1$. Check that 2 is a primitive root of 3, 5, 11, and 13 and not a primitive root of 7 and 17. Prove the following: I_{2n} is a single cycle if and only if $2n + 1$ is a prime and 2 is a primitive root of $2n + 1$. (Hint: To start the proof, use Exercise 1 to argue that $2n + 1$ must be prime.)

7. Invent an interlacing shuffle that divides the deck into three equal parts and study it. Hint: For 12 cards, the permutation is

$$\begin{pmatrix} 9 & 5 & 1 & 10 & 6 & 2 & 11 & 7 & 3 & 12 & 8 & 4 \\ 1 & 2 & 3 & 4 & 5 & 6 & 7 & 8 & 9 & 10 & 11 & 12 \end{pmatrix}$$

or

$$\begin{pmatrix} 1 & 2 & 3 & 4 & 5 & 6 & 7 & 8 & 9 & 10 & 11 & 12 \\ 3 & 6 & 9 & 12 & 2 & 5 & 8 & 11 & 1 & 4 & 7 & 10 \end{pmatrix}$$

5. The Josephus Permutation

There is an amusing legend about Flavius Josephus, a famous historian who lived in the first century A.D.

The story goes as follows. In the Jewish revolt against Rome, Josephus and 39 of his comrades were holding out against the Romans in a cave. With defeat imminent, they resolved that, like the rebels at Masada, they would rather die than be slaves to the Romans. They decided to arrange themselves in a circle. One man was designated as number one, and they proceeded clockwise around the circle of 40 men, killing every 7th man (the numbers 40 and 7 are different in other versions of the tale). Now it is obvious whose turn it was to be killed on the first time around: numbers 7, 14, 21, 28, and 35. From this point on, the outcome is not as clear. The next number would be 42; however, this has to be reduced modulo 40 to 2, and man number 2 is indeed the next to go. But the next after that is not man number 9. We are counting men, not seats in the circle. Seat number 7 is unoccupied. So the next man to be slaughtered is number 10.

Let us interrupt at this point, in order to finish telling the story. Josephus (according to the story) was among other things an accomplished mathematician; so he instantly figured out where he ought to sit in order to be the last to go. (The reader might like to try Exercise 1 at this point.) But when the time came, instead of killing himself he joined the Roman side. And so he lived to write his famous histories: *The Antiquities* and *The Jewish War*.

We are going to take the point of view that the Josephus "game" has resulted in a permutation of the 40 men. Thus the 7th man is now first, the

14th man is now second, and so on. Actually, it will be a little more convenient to discuss instead the inverse permutation, which has the form

$$\begin{pmatrix} 1 & 2 & 3 & 4 & 5\,6 & 7\,8\,9 \cdots 40 \\ 7 & 14 & 21 & 28 & 35\,2 & 10\,?\,? \cdots \ ? \end{pmatrix}$$

Let us generalize. Instead of 40 men in a circle, let us have n. Instead of liquidating every 7th man, let it be every dth. We shall write $J_{n,d}$ for the resulting permutation, and we might call it the *Josephus permutation* for the numbers n and d. We have

$$J_{n,d} = \begin{pmatrix} 1 & 2 & 3 & 4 \cdots n \\ d & 2d & \cdots & ? \end{pmatrix}$$

where the bottom row has multiples of d until n is reached, followed by mysterious entries we have not mastered. In the story above, the permutation that took place was $J_{40,7}$.

An extensive account of the Josephus permutation appears in Chapter 15 (pp. 286–301) of the book *Mathematische Unterhaltungen und Spiele* (translation: "Mathematical Recreations and Games") by W. Ahrens (Teubner: Leipzig, 1901). Ahrens gives a number of references to earlier papers on the subject that appeared in various journals. (All of these references are in German or French.) These authors were largely concerned with the aspect of the Josephus permutation that dominates the original story; they sought a formula for the number appearing at the end of the bottom row of the permutation (the "last man to be killed"). More generally, they sought a formula for the number appearing in the rth position for any integer r between 1 and n; and indeed they found an interesting formula of this kind. However, they did not raise the questions that shall occupy our attention: What is the order of the Josephus permutation, and what is its cycle structure?

There seems to be no substantial literature on the Josephus permutation subsequent to the Ahrens book; the references we have found are listed at the end of this section.

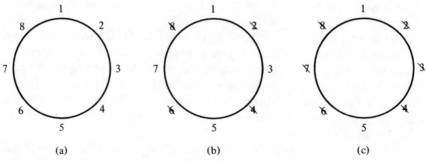

(a)   (b)   (c)

Figure 3.1

We shall now compute the cycle structure of the Josephus permutation in an example. At present we shall consider only the case $d = 2$; so, instead of $J_{n,2}$ we simplify by writing J_n.

Let us envisage the computation of J_8 as in Figure 3.1 We put the numbers 1 to 8 in a circle, as shown in part (a). On the first round the even numbers 2, 4, 6, 8 are deleted in succession. In part (b) they have been crossed out, and we tackle the odd numbers for the second round. The next to go are 3 and 7, as shown in part (c). We finish with 5 and 1 in that order.* All in all, we have

$$J_8 = \begin{pmatrix} 1\ 2\ 3\ 4\ 5\ 6\ 7\ 8 \\ 2\ 4\ 6\ 8\ 3\ 7\ 5\ 1 \end{pmatrix}$$

Here is the expression of J_8 as a product of disjoint cycles:

$$J_8 = (1248)(3675)$$

The order of J_8 is 4.

Table 3.1 exhibits the cycle structure of J_n for all n up to 18. At first glance, the table will undoubtedly present an impression of hopeless chaos.

Table 3.1

n	J_n
1	(1)
2	(1 2)
3	(1 2)(3)
4	(1 2 4)(3)
5	(1 2 4 5 3)
6	(1 2 4 3 6 5)
7	(1 2 4)(3 6)(5)(7)
8	(1 2 4 8)(3 6 7 5)
9	(1 2 4 8 7 9 3 6 5)
10	(1 2 4 8)(3 6)(5 10)(7)(9)
11	(1 2 4 8 9 3 6)(5 10 11 7)
12	(1 2 4 8 7 3 6 12 9 11)(5 10)
13	(1 2 4 8 5 10 13 11 7)(3 6 12)(9)
14	(1 2 4 8 3 6 12 9 7 14 13 5 10 11)
15	(1 2 4 8)(3 6 12)(5 10 9)(7 14)(11 13)(15)
16	(1 2 4 8 16)(3 6 12 15 9)(5 10 7 14 13)(11)
17	(1 2 4 8 16 11 9)(3 6 12 13 17)(5 10)(7 14)(15)
18	(1 2 4 8 16 17 13 15 9 18 5 10 3 6 12 11 7 14)

*Instead of knocking down imaginary rebels in a circle, the reader might prefer to achieve the Josephus permutation by using cards. For instance, take cards numbered from 1 to 8. Holding them in that order, with 1 on top, place the top card on the bottom, the next card on the table, the next card on the bottom, the next card on the table, and continue until the entire pile has been moved to the table. The cards will emerge in the order given by J_8: 2, 4, 6, 8, 3, 7, 5, 1.

Nevertheless there is a lot of structure present. Some facts are stated below, without proof. When this manuscript was first written in early 1972, nearly all of these statements were listed as conjectures, with extensive numerical evidence to support them. Now they are theorems, thanks largely to Andrew Gallant and Michael O'Nan. The proofs of these theorems (except for the very last which concerns class numbers) are not terribly hard (admittedly this is Monday morning quarterbacking, since for many months no proofs were available at all). But it would take up too much space, and be somewhat out of proportion, to present the proofs here. An ambitious reader can look at them as challenging exercises supplementing the easier ones at the end of the section.

1. If n is a power of 2, say $n = 2^r$, then the order of J_n is $r + 1$. Thus the order of J_4 is 3, the order of J_8 is 4, and the order of J_{16} is 5 (these three assertions are evident from Table 3.1). Beyond the range of the table, the order of J_{32} is 6, the order of J_{64} is 7, and so forth.
2. Again take $n = 2^r$. If $r + 1$ is an odd prime (covering the values $r = 2, 4, 6,$ and 10 in Table 3.2), J_n consists of a fixed point and $(r + 1)$-cycles. The remaining entries in Table 3.2 suggest that J_{2^r} contains an enormous number of $(r + 1)$-cycles together with a tiny supplement of shorter cycles. The nature of the supplement can in fact be described exactly, but we shall not do so here; it is a little complicated.

Table 3.2

r	$n = 2^r$	Structure of J_n
2	4	A 3-cycle and a fixed point
3	8	2 4-cycles
4	16	3 5-cycles and a fixed point
5	32	5 6-cycles and a transposition
6	64	9 7-cycles and a fixed point
7	128	16 8-cycles
8	256	28 9-cycles, a 3-cycle, and a fixed point
9	512	51 10-cycles and a transposition
10	1024	93 11-cycles and a fixed point

3. Suppose n has the form $2^r - 1$. Then $(124 \cdots 2^{r-1})$ is an r-cycle in J_n. All other cycles have length less than r, and all such possible lengths appear. For an illustration, see J_{15} in Table 3.1.
4. Suppose that n and $2n + 1$ are both prime and that $n \equiv 1 \bmod 4$. Then J_n is an n-cycle.

In Table 3.1 the entry $n = 5$ is an illustration. The next value is $n = 29$.

5. There is a sharper theorem that is stated in terms of the notion of a primitive root (compare Exercise 6 of the preceding section, and note the resemblance between that exercise and the present result). J_n is an n-cycle if and only if $2n + 1$ is a prime and 2 is a primitive root of $2n + 1$.

The entries $n = 2, 5, 6, 9, 14,$ and 18 in Table 3.1 serve as illustrations.

6. If n and $2n + 1$ are both prime and $n \equiv 3 \bmod 4$, then J_n consists of two cycles.

Table 3.3 lists the first nine values of n to which this applies.

Table 3.3

n	Length of cycles in J_n
3	2, 1
11	7, 4
23	14, 9
83	47, 36
131	72, 59
179	99, 80
191	104, 87
239	132, 107
251	136, 115

For several months we were mystified. What, if anything, did the lengths of these cycles mean? It finally turned out that there was a surprising connection with number theory.

Form the ring, say R, of all numbers of the form $a + b\sqrt{-(2n + 1)}$, with a, b running over the integers. For instance, with $n = 23$, R is the set of all numbers $a + b\sqrt{-47}$. Now it is a fact that (except for the top entry of Table 3.3, corresponding to $\sqrt{-7}$) unique factorization into primes does not hold in R. In an advanced portion of number theory, called algebraic number theory, such a failure of unique factorization is studied closely. It is useful to attach to the ring a numerical invariant which, so to speak, measures the failure of uniqueness (the bigger the invariant, the worse the failure). The invariant is called the *class number* of R. Now we can describe the meaning of the numbers in Table 3.3: The difference of the lengths of the cycles is the class number of R. Thus, for instance, $5 = 14 - 9$ is the class number of the ring of numbers $a + b\sqrt{-47}$, and $25 = 132 - 107$ is the class number of the ring of numbers $a + b\sqrt{-479}$.

All of this discussion concerned the Josephus permutation $J_n = J_{n,2}$. As of this writing, very little is known concerning $J_{n,d}$ with $d > 2$.

Professors C. D. Feustel, L. Johnson, and J. Layman of Virginia Polytechnic Institute have made some computations on the VPI computer.

They have tabulated the cycle structure of $J_{n,3}$ for $n \leq 1000$. Also, they found the following information concerning the possibility of $J_{n,d}$ being a cycle:

For $n \leq 2155$, $J_{n,3}$ is a cycle if and only if $n = 3, 5, 27, 89, 1139,$ 1219, 1921, or 2155.

For $n \leq 1000$, $J_{n,4}$ is a cycle if and only if $n = 5, 10, 369,$ or 609.

For $n \leq 300$, $J_{n,7}$ is a cycle if and only if $n = 11, 21, 35, 85, 103,$ 161, or 231.

The results for $d = 3$ and $d = 7$ suggest that if d is odd and $J_{n,d}$ is a cycle, then n must be odd. This is in fact true, and it can be deduced from Exercise 7(c).

Exercises

1. What seat did Josephus occupy in order to be the last to go? In other words, what is $J_{40,7}(40)$?
2. Prove that $J_{n+1,n}$ has 1 as a fixed point.
3. Prove that both $J_{n,n}$ and $J_{n,2n-1}$ have 3 as a fixed point.
4. (a) Prove that $J_{2n-2,n}$ has 2 as a fixed point.
   (b) Prove that $J_{3n-3,n}$ has 3 as a fixed point.
   (c) Can you generalize parts (a) and (b)?
5. Prove that $J_{n^2-1,n}$ contains the transpostion $(1n)$.
6. Let $n = 2^r - 1$. Verify that J_n leaves n fixed for $r \leq 6$ (for $r \leq 4$, see Table 3.1). (Remark: The result is true for any r.)
7. (a) Prove that

   $$J_n = (n - 1, n)(n - 2, n - 1, n) \cdots (2\ 3 \cdots n)(1\ 2\ 3 \cdots n)$$

   (b) More generally, prove that

   $$J_{n,d} = (n - 1, n)^{d-1}(n - 2, n - 1, n)^{d-1} \cdots (2\ 3 \cdots n)^{d-1}$$
   $$(1\ 2\ 3 \cdots n)^{d-1}$$

   (c) By using part (b), or otherwise, prove the following:
      (i) If d is odd, $J_{n,d}$ is an even permutation.
      (ii) If d is even, $J_{n,d}$ is an even permutation if n is congruent to 0 or 1 mod 4, and an odd permutation if n is congruent to 2 or 3 mod 4.
8. Write the product that occurred in Exercise 7(a) in the reverse order. If we call the result σ, we have

   $$\sigma = (1\ 2 \cdots n)(2\ 3 \cdots n) \cdots (n - 2, n - 1, n)(n - 1, n)$$

   If n is even, prove that σ is a product of $n/2$ disjoint transpositions; if n is odd, prove that σ is a product of $(n - 1)/2$ disjoint transpositions and a fixed point.
9. How many fixed points does $J_{8,9}$ have? (At present, this is an

isolated curiosity with an unusually large number of fixed points. It was noted by Michael O'Nan.)

10. Prove that $J_{3,5}$ is a cycle. What is the next value of n such that $J_{n,5}$ is a cycle?

11. (a) Let k be the least common multiple of the integers from 1 to n. Prove that $J_{n,d}$ depends only on the residue of d mod k.

    (b) Prove that $J_{n,d}$ is the identity if and only if d is congruent to 1 mod k.

    (c) Describe $J_{n,d}$ if d is divisible by k.

Bibliography

In addition to Ahrens, and the papers cited by Ahrens, we have located the following references to the Josephus permutation:*

D. J. Aulicino and Morris Goldfeld, "A new relation between primitive roots and permutations," *American Mathematical Monthly* 76 (1969): pp. 664–666. (Our Ex. 7 identifies the permutation they study with the Josephus permutation.)

W. W. R. Ball, *Mathematical Recreations and Essays,* 11th ed., revised by H. S. M. Coxeter (New York: Macmillan, 1939), pp. 32–36. Ball gives several earlier references, including one to Hegesippus for the Josephus story itself. (Hegesippus was the name attached to a free Latin translation of Josephus made in the fourth century. Josephus himself wrote in Greek.)

A. P. Domoryad, *Mathematical Games and Pastimes* published in Russian in 1961, Halina Moss, trans. (New York: Macmillan, 1964). Section 21 (pp. 124–126) concerns the Josephus permutation.

Howard W. Eves, *Mathematical Circles Revisited* (Boston: Prindle, Weber, and Schmidt, 1971), pp. 71–72.

Martin Gardner, *The Unexpected Hanging* (New York: Simon & Schuster, 1969), pp. 155–158.

F. Jakóbczyk, "On the generalized Josephus problem," *Glasgow Math. J.* 14 (1973): pp. 168–173.

D. E. Knuth, *The Art of Computer Programming* (Reading, Mass.: Addison-Wesley), vol. I *Fundamental Algorithms,* 1968, Ex. 22, p. 158; vol. III *Sorting and Searching,* Ex. 2, pp. 18–19. See also p. 181 of the second edition of vol. I.

M. Kraitchik, *Mathematical Recreations* (New York: Norton, 1942), pp. 93–94.

N. S. Mendelsohn, Problem E 898, *American Mathematical Monthly* 57 (1950): pp. 34–35; solution, pp. 488–489.

*We are indebted to Joshua Barlaz, Martin Gardner, and Donald Knuth for help with these references. Thanks also to Donald Knuth for telling us about a Slavonic version of the Josephus book in which the following phrase occurs:

. . . he counted the numbers with cunning and thereby misled them all.

This comes a lot closer to making Josephus a mathematician than the standard Greek version which says at this point:

. . . he however (should one say by fortune, or by the providence of God?) was left alone with one other . . .

James R. Newman, *The World of Mathematics,* vol. 4 (New York: Simon & Schuster, 1956), pp. 2428–2429.

R. A. Rankin (R. A. MacFhraing), "Aireamh muinntir Fhinn is Dhubhain, agus sgeul Isoephuis dà fhichead Iudhaich," *Proc. Roy. Irish Acad. Sect. A.* 52 (1948): pp. 87–93.

W. J. Robinson, "The Josephus Problem," *Math. Gazette* 44 (1960): pp. 47–52.

F. Schuh, *The Master Book of Mathematical Recreations,* F. Göbel, trans. (New York: Dover, 1968); pp. 372–375 deal with the Josephus permutation, although not by that name.

Chapter 4

Group Theory

A strong characteristic of the mathematics of the last hundred years or so has been its tendency to go abstract. To a nonmathematician this statement may seem ludicrous; after all, he could say, mathematics is already totally abstract. There is no doubt that all mathematics is abstract —even something so basic as what is an integer is already an abstraction to a highly nontrivial degree—however, in the framework of mathematics we can have diverse levels of abstractness.

This characteristic will be well illustrated in what we are about to do in this chapter. We have just studied something about the permutations of a set of objects. Looking at the properties enjoyed by permutations we make a long jump, extracting what formal properties seem essential in that setup of permutations, to define a certain formal algebraic system called a *group*. It will be defined so as to have the set of permutations of a set as a special model. Thus, anything we say about *all* groups, will hold true for the *particular* example of the symmetric group.

It may seem strange to expect to say more about a broader situation than about a particular case of it. However, this is often true. In freeing ourselves from the particularity of a situation, by stripping it of its inessentials, we are often in a better position to view it more clearly. This is one reason for abstracting in mathematics. Its great justification has been the success that this abstract method has enjoyed, especially in clarifying and answering many classical questions.

1. Definition and Examples of Groups

We being immediately with the definition of a group. Given a set S, by a *product* or *operation* on S, we shall mean a rule for combining two of the elements of S to obtain again an element of S.

Definition. A nonempty set G is said to be a *group* if in G there is defined an operation $\cdot$ such that:

1. a, b ε G implies $a \cdot b$ ε G.
2. If a, b, c ε G, then $(a \cdot b) \cdot c = a \cdot (b \cdot c)$. This is called the *associative* law.
3. There exists an element e ε G such that $a \cdot e = e \cdot a = a$ for all a ε G. We describe this by saying that G has an *identity element*.
4. For every a ε G there exists an element a^{-1} ε G such that $a \cdot a^{-1} = a^{-1} \cdot a = e$. This is described by saying that G has *inverses*.

Note that in our definition of a group, the first axiom, namely, a, b ε G forces $a \cdot b$ ε G, is superfluous if by operation we mean what we wrote above. However, we include it to emphasize that G is *closed* (this is the technical term used) under the operation.

To be honest, we were almost at the general concept of a group in Section 2 of Chapter 2. There, when we spoke about *abelian groups,* except for our insisting on commutativity and the fact that the operation was written as $+$, we had already reached the concept.

In this chapter we are concerned with all groups, commutative or not. Because commutativity makes groups behave a lot better, mathematicians have singled them out and have given them the name that we have already used.

Definition. A group G is said to be *abelian* or *commutative* if for every a and b in G, $a \cdot b = b \cdot a$.

> The word *abelian* comes from the name of the Norwegian mathematician Nils Henrik Abel (1802–1829), who made many contributions in mathematics, but is perhaps best known for having shown that a general polynomial of degree five or greater has no solutions in terms of radicals. He is probably the greatest Norwegian scientist.

EXAMPLE 1. Let G be the set of integers and let, for a, b ε G, $a \cdot b$ be $a + b$. G is a group under this operation. The integer 0 acts as the e; for a^{-1} (relative to this operation) we use $-a$.

EXAMPLE 2.   Let G be the set consisting of the integers 1, -1 with the operation $\cdot$ the ordinary multiplication of integers. It is readily verified that G is a group having two elements.

EXAMPLE 3.   Let S be any set and let $G = A(S)$, the set of all one-to-one mappings of S onto itself. For τ, $\sigma \ \varepsilon \ G$ let their product $\sigma \cdot \tau$ be defined by the composition of functions, as used in the preceding chapter. We saw earlier that G is a group under this operation. In fact this group, in the special case when S has a finite number of elements, was called the symmetric group or group of permutations. When S has n elements we write G as S_n; note that in S_3, the elements $\sigma = (123)$ and $\tau = (12)$ do not commute; that is, $\sigma \cdot \tau \neq \tau \cdot \sigma$, for $\sigma \cdot \tau = (13)$ while $\tau \cdot \sigma = (23)$. Thus S_3 is an example of a non-abelian group.

EXAMPLE 4.   Let $n > 1$ be an integer and let $G = Z_n$ be the integers modulo n under addition. Our discussion in Chapter 2 showed that G is an abelian group having n elements.

EXAMPLE 5.   Let p be a prime number and let G be the nonzero integers modulo p under multiplication. Our discussion in Chapter 2 revealed that G is a group.

We examine Example 5 a little more closely. Recall that $[a] = [b]$ if and only if $a \equiv b \bmod p$. Let $G = \{[1], [2], \ldots, [p-1]\}$; in G we define $[a] \cdot [b] = [ab]$. Our work in number theory showed that G is a group under this operation.

Let us take the special case $p = 5$; G consists of the elements $[1]$, $[2]$, $[3]$, $[4]$. What is the *multiplication table* of G? We write it as shown in Table 4.1.

Table 4.1

	[1]	[2]	[3]	[4]
[1]	[1]	[2]	[3]	[4]
[2]	[2]	[4]	[1]	[3]
[3]	[3]	[1]	[4]	[2]
[4]	[4]	[3]	[2]	[1]

What does this bunch of symbols represent? To read off $[i] \cdot [j]$ look at the entry in the $[i]$ row and $[j]$ column; thus for instance, with $i = 3$ and $j = 4$ we get $[3] \cdot [4] = [2]$.

We look even further at the example. Let $a = [3]$. Even though we have not formally defined *powers* in groups we can consider

$$a^2 = [3] \cdot [3] = [4], \quad a^3 = ([3] \cdot [3]) \cdot [3] = [2], \quad a^4 = (([3] \cdot [3]) \cdot [3]) \cdot [3] = [1]$$

Note that here the powers of a sweep out all of G; that is, given $[i] \neq [0]$ then $[i] = a^k$ for some k. This is a particular example of a *cyclic group*, a type of group which we define more formally below.

EXAMPLE 6. Let $n > 1$ be an integer. Let G consist of the symbols $\{e, a, a^2, \ldots, a^{n-1}\}$ where we declare:

$$a^i \cdot a^j = a^{i+j} \quad \text{if} \quad i+j < n$$
$$a^i \cdot a^j = a^{(i+j)-n} \quad \text{if} \quad i+j > n$$
$$a^i \cdot a^{n-i} = e$$

It can be verified that G is an abelian group having n elements; its identity element is e and the inverse of an element $a^i \neq e$ is defined by $(a^i)^{-1} = a^{n-i}$. G is called a *cyclic group* of order n.

Note that Example 4 is a cyclic group of order n. Let us take a look at that. Let $G = Z_n$, the integers modulo n under addition. What does this mean? $G = \{[0], [1], \ldots, [n-1]\}$ where $[a] + [b] = [a+b]$ and where $[x] = [y]$ if and only if $x \equiv y \bmod n$. If $a = [1]$ then the powers of a—since the operation in G is addition this really means the multiples of a—sweep out G.

Exercises

1. Verify that Example 2 is indeed a group.
2. Are the following systems groups?
   (a) G = the set of integers, $a \cdot b = a - b$.
   (b) $G = \{\sigma \varepsilon A(S) | \sigma(s_0) = s_0\}$, s_0 is a fixed element in S.
   (c) G = the set of all pairs (a, b) with $a, b \varepsilon Z$ where $(a, b) \cdot (c, d) = (a + c, b + d)$.
3. By the group table of the group G having n elements $g_1, \ldots, g_n$ we mean the square array

	g_1	g_2	g_3	$\cdots$	g_n
g_1					
g_2					
g_3					
$\vdots$					
g_n					

   where the entry in the ith row, jth column is $g_i \cdot g_j$.
   (a) Write the group table for $G = Z_3$.
   (b) Write the group table for $G = S_3$.

4. If the group G is abelian what can you say about its group table?
5. Show that in the group table of a group G we find every element of G once and only once in each row and in each column.
6. Is

	a	b	c	d
a	a	d	c	b
b	b	a	d	c
c	c	b	a	d
d	d	c	b	a

the group table of some group?
7. Let σ be the rotation of the plane S through a counterclockwise angle of 90° and let τ be the reflection of the plane about the y-axis. These are one-to-one mappings of the plane S onto itself. Show that in $A(S)$
   (a) $\sigma^4 = i$
   (b) $\tau^2 = i$
   (c) $\tau\sigma\tau = \sigma^{-1}$
   (d) Show that the set $\{i, \sigma, \sigma^2, \sigma^3, \tau, \tau\sigma, \tau\sigma^2, \tau\sigma^3\}$ is a non-abelian group.

2. Some Beginning Notions and Results

Before going much further we need a little language and some basic facts that will facilitate our computing in groups. This will be our first objective.

Let G be a group and let $a \, \varepsilon \, G$. We define the powers of a as follows:

$$a^0 = e$$
$$a^1 = a$$
$$a^2 = a \cdot a$$
$$\cdot$$
$$\cdot$$
$$\cdot$$
$$a^k = (a^{k-1}) \cdot a, \qquad \text{for any positive integer } k$$

For the negative powers we use $a^{-n} = (a^{-1})^n$ where $n > 0$. In this there is a little vagueness as to meaning, for as yet, we do not know that a^{-1} is unique. This will be resolved shortly.

The following basic facts concerning powers can be proved by induction; the details are tedious and we omit them.

1. $a^i \cdot a^j = a^{i+j}$
2. $(a^i)^j = a^{ij}$

for all integers i and j (positive, negative, or zero).

If G should have a finite number of elements, we call G a *finite group* and call the number of elements in G the *order* of G. We denote the order of G by $o(G)$.

We proceed to our first results about groups. In fact, Lemmas 4.1 and 4.2 were essentially proved in Chapter 2 in Lemma 2.1 and its corollary. Because the setting there was commutative, and commutativity is not being assumed here, for the sake of completeness we redo the proof.

Lemma 4.1.   In a group G the unit element is unique. Moreover, the inverse of an element in G is also unique. If $a \, \varepsilon \, G$, then $(a^{-1})^{-1} = a$. Finally, for $a, b \, \varepsilon \, G$, $(a \cdot b)^{-1} = b^{-1} \cdot a^{-1}$.

Proof.   What is it that we must prove? To show uniqueness of e we must show that if $a \cdot f = f \cdot a = a$ for all $a \, \varepsilon \, G$, then $f = e$.

Suppose then that $a \cdot f = f \cdot a = a$ for all $a \, \varepsilon \, G$; then in particular using $a = e$, $e \cdot f = f \cdot e = e$. However, because e is an identity for G, $e \cdot f = f$. The net result of this is that $e = f$.

Now to the uniqueness of inverses. Suppose $a \cdot b = e$ and $a \cdot c = e$. Now we know that there is an $x \, \varepsilon \, G$ such that $x \cdot a = e$. This yields

$$x \cdot (a \cdot b) = x \cdot e = x \cdot (a \cdot c) = (x \cdot a) \cdot c = e \cdot c = c,$$
$$(x \cdot a) \cdot b = e \cdot b = b$$

and since

$$x \cdot (a \cdot b) = (x \cdot a) \cdot b$$

we read that $b = c$. We have shown that every element in G has a unique inverse in G.

If $a \, \varepsilon \, G$, then by definition $a^{-1} \cdot a = a \cdot a^{-1} = e$ and $(a^{-1}) \cdot (a^{-1})^{-1} = e$. Thus from the uniqueness of the inverse of a^{-1} in G we get $(a^{-1})^{-1} = a$.

To see that $(a \cdot b)^{-1} = b^{-1} \cdot a^{-1}$ notice that

$$(a \cdot b) \cdot (b^{-1} \cdot a^{-1}) = ((a \cdot b) \cdot b^{-1}) \cdot a^{-1} = (a \cdot e) \cdot a^{-1} = a \cdot a^{-1} = e$$

The very definition of inverse then tells us that $(a \cdot b)^{-1} = b^{-1} \cdot a^{-1}$.

In the course of proving the uniqueness of the inverse of an element in G we actually proved something more. This is the possibility of cancellation in relations involving group elements; to be more precise, we claim that if $a \cdot b = a \cdot c$, then $b = c$ and that if $d \cdot a = g \cdot a$, then $d = g$. This is our next goal.

Lemma 4.2.   In a group G we can cancel from the *same* side of a relation; that is:

1. If $a \cdot b = a \cdot c$ with $a, b, c, \, \varepsilon \, G$, then $b = c$.
2. If $d \cdot a = g \cdot a$ with $a, d, g \, \varepsilon \, G$, then $d = g$.

Proof.   Suppose that $a \cdot b = a \cdot c$; therefore $a^{-1} \cdot (a \cdot b) = a^{-1} \cdot (a \cdot c)$. Using the associative law, we get $(a^{-1} \cdot a) \cdot b = (a^{-1} \cdot a) \cdot c$; that is, $e \cdot b = e \cdot c$. But $b = e \cdot b = e \cdot c = c$. Therefore $b = c$. We similarly prove the other result.

Consider now a finite group G having n elements. If $a \, \varepsilon \, G$, consider the elements $e, a, a^2, \ldots, a^n$; these $n + 1$ elements are all in G, which has only n elements. By the Pigeon-Hole Principle, two of these must be equal. Stated more formally, there must be integers i and j with $0 \leq i < j \leq n$ such that $a^i = a^j$. Since $j > i$, $j = i + k$ where $k > 0$. Therefore $a^i \cdot e = a^{i+k} = a^i \cdot a^k$; by Lemma 4.2 we can cancel the a^i to get $a^k = e$. Note that $n \geq k > 0$, and that k depends on a. We have proved

Lemma 4.3.   In a finite group G of order n every element satisfies $a^k = e$ for some positive integer k, $0 < k \leq n$, which depends on a.

We want more precise information about k. Can we have an exponent k that works for all elements? Clearly yes!

Let $a \, \varepsilon \, G$, G of order n, and suppose that $a^k = e$ where $0 < k \leq n$. Thus $k \mid n!$; hence we can write $n! = kt$ where t is an integer. In consequence

$$a^{n!} = a^{kt} = (a^k)^t = e^t = e$$

We have shown the

Corollary.   If G is of order n, then $a^{n!} = e$ for all $a \, \varepsilon \, G$.

We shall presently see that we can get a sharper result, namely, that $a^n = e$. We leave this until later.

Definition.   An element $a \, \varepsilon \, G$ is said to be of *finite order* if $a^k = e$ for some *positive* integer k.

It may very well happen that there are elements in G which are not of finite order. For instance, if G is the group of integers under addition, then G has no nonzero elements of finite order.

Let $a \, \varepsilon \, G$ be of finite order. By the *order* of a, written $o(a)$, we shall mean the *smallest positive* integer u such that $a^u = e$.

Note that Lemma 4.3 assures us that in a finite group every element is of finite order.

We close this section with a result about an element of finite order.

Lemma 4.4.   Let $a \, \varepsilon \, G$ be of finite order $o(a)$ and suppose that $a^k = e$ where $k > 0$. Then $o(a) \mid k$.

Proof.   By the definition of $o(a)$, since $a^k = e$, $k \geq o(a)$. Using the Euclidean algorithm we can write $k = o(a)m + r$ where $0 \leq r < o(a)$.

Thus

$$e = a^k = a^{o(a)m + r} = (a^{o(a)})^m \cdot a^r = e^m \cdot a^r = a^r$$

Since $0 \leq r < o(a)$ and $a^r = e$, r cannot be positive by the very definition of $o(a)$, hence we must have $r = 0$. Therefore $k = o(a)m$ and $o(a)|k$.

Exercises

1. What do the rules of exponents become in case G is the group of integers under addition?
2. Give an example of a group in which $a \cdot b = c \cdot a$ yet $b \neq c$. (This clearly indicates you may only cancel from the *same* side of an equation.)
3. If every element in a group G satisfies $x^2 = e$ prove that G must be abelian.
4. Let G be a group. Prove that G is abelian if and only if for all $a, b \ \varepsilon \ G$ and all integers i, $(a \cdot b)^i = a^i \cdot b^i$.
5. Suppose in a group G we have $(a \cdot b)^i = a^i \cdot b^i$ for three consecutive integers i. Prove $a \cdot b = b \cdot a$.
6. Let G be a group. We define for $a, b \ \varepsilon \ G$, $a \sim b$ if there is some $x \ \varepsilon \ G$ such that $b = x^{-1} \cdot a \cdot x$. Prove that this defines an equivalence relation on G.
7. Give the orders of all elements in S_3, the symmetric group of degree 3. Do the same for S_4.
8. Let G be a finite set in which an associative product is defined such that G is closed relative to this product. Suppose that both cancellation laws hold in G; that is, $a \cdot b = a \cdot c$ forces $b = c$ and $x \cdot a = y \cdot a$ forces $x = y$. Prove that G is a group.

3. Subgroups

If G is a group, then the subsets of G which should be of most interest to us are those that reflect the algebraic structure carried by G. This quickly leads us to the notion of a subgroup of G.

Definition. If G is a group and H is a nonempty subset of G, then H is said to be a *subgroup* of G if, relative to the operation of G, H itself forms a group.

There are many ways of giving examples of subgroups of groups. One is, clearly, to pick some specific group and to exhibit subgroups thereof. Another is to give general constructions, in any group, of subgroups. We shall do a little of both. We begin with the first type. Note, however, that any group G always has two trivial subgroups, namely, G itself and the set $\{e\}$.

EXAMPLE 1.   Let G be the group of integers under addition, and let H be the subset consisting of the even integers. Then H, as is immediately verified, is a subgroup of G.

EXAMPLE 2.   Let G be the group of integers under addition; let $n > 1$ be an integer. Define H_n to be all multiples of n. We leave it to the reader to verify that H_n is a subgroup of G. Note that Example 1 is the special case of Example 2 in which $n = 2$.

EXAMPLE 3.   Let $G = S_3$, the symmetric group of degree 3, and let $H = \{e, (123), (132)\}$. We leave it to the reader to verify that H is a subgroup of G.

EXAMPLE 4.   Let S be a nonempty set and let $G = A(S)$ be the set of all one-to-one mappings of S onto itself. Pick $s_o \in S$ and let $H = \{\sigma \in A(S)|\sigma(s_o) = s_o\}$. We claim that H is a subgroup of G. To see this we check that the group axioms hold for H.

(1) If $\sigma, \tau \in H$, then $\sigma(s_o) = s_o$, $\tau(s_o) = s_o$ hence $(\sigma \cdot \tau)(s_o) = \tau(\sigma(s_o)) = \tau(s_o) = s_o$, in consequence of which $\sigma \cdot \tau \in H$.
(2) If $\sigma, \tau, \gamma \in H$ then, being in G, $\sigma \cdot (\tau \cdot \gamma) = (\sigma \cdot \tau) \cdot \gamma$; therefore the associative law holds in H.
(3) If i is the identity element of G, then $i(s) = s$ for all $s \in S$ hence, in particular, $i(s_o) = s_o$. This places $i \in H$.
(4) If $\sigma \in H$, then $\sigma(s_o) = s_o$ from which we get $\sigma^{-1}(\sigma(s_o)) = \sigma^{-1}(s_o)$. But $\sigma^{-1}(\sigma(s_o)) = (\sigma \cdot \sigma^{-1})(s_o) = i(s_o) = s_o$. The upshot of this is that $\sigma^{-1}(s_o) = s_o$ whence $\sigma^{-1} \in H$.

Having verified that H satisfies all the group axioms relative to the operation of G we have that H is a subgroup of G.

EXAMPLE 5.   Let $G = \{[1], [2], [3], [4], [5], [6]\}$ be the set of nonzero integers modulo 7 relative to multiplication. Let $H = \{[1], [2], [4]\}$; a simple check reveals that H is a subgroup of G. If $K = \{[1], [6]\}$, it is also easy to see that K is a subgroup of G.

Having seen some subgroups of some specific groups we now go on to a discussion of subgroups of general groups. But first we need a criterion that a subset of a group be a subgroup. This is

Lemma 4.5.   A nonempty subset, H, of a group G is a subgroup if and only if

1. $a, b \in H$ implies $a \cdot b \in H$ (i.e., H is closed relative to the operation of G).
2. $a \in H$ implies $a^{-1} \in H$.

Proof. To begin with, when H is a subgroup of G, from the very group axioms, it follows that H is closed under the multiplication of G and for every $a \in H$, a^{-1} must also be in H.

Suppose now that H is a subset of G for which $a, b \in H$ implies $a \cdot b \in H$ and $a^{-1} \in H$. Clearly the multiplication in H is associative because we already know it to be associative in all of G. To show that H is a subgroup we merely need that e is in H. Because H is nonempty there is an element $a \in H$; hence $a^{-1} \in H$ by assumption. Therefore $e = a \cdot a^{-1}$ is also in H since H is closed relative to the product. This finishes the proof.

In the special case that H, in addition, has a finite number of elements, the situation becomes even simpler. Here, closure is enough. We prove

Theorem 4.1. Let G be a group and H a nonempty finite subset such that $a, b \in H$ implies $a \cdot b \in H$. Then H is a subgroup of G.

Proof. According to Lemma 4.5, we must establish that, in addition to closure (which is already given), H contains the inverses of all its elements. Let $a \in H$; if $a = e$, then $a^{-1} = e$ so it, too, would be in H. Assume $a \neq e$. Look at all the positive powers of a. They lie in H, which is a finite set. So two of them must be equal; that is, $a^r = a^s$ for some $r < s$. We are allowed to cancel in a group, so we get $a^{s-r} = e$. If $s - r$ happens to be 1, then $a = e$, something we have already ruled out. So we can suppose $s - r > 1$. But then $b = a^{s-r-1}$ is a positive power of a, hence lies in H. Since $a \cdot b = a^{s-r} = e$, b is the required inverse of a in H.

It is a little cumbersome to keep writing $a \cdot b$ for the product of a and b in G. *Henceforth we drop the dot and write $a \cdot b$ as ab.*

Let G be a group and let $a \in G$. Let $H = \{a^i | i = 0, \pm 1, \pm 2, \ldots\}$. It is fairly easy to verify that H is a subgroup of G. H is called the subgroup *generated* by a. We call a group G a *cyclic group* if, for some element $a \in G$, G consists of the positive, zero, and negative powers of a. The subgroup H just constructed is a cyclic subgroup of G.

Thus we have a general method of generating some subgroups, albeit very special ones, in groups. What can one say about a group G if G has no subgroups other than the trivial ones G and $\{e\}$? This is

Theorem 4.2. If a group $G \neq \{e\}$ has no subgroups other than $\{e\}$ and itself, then G is a cyclic group of prime order p.

Proof. Since $G \neq \{e\}$ by assumption, there is an element $a \neq e$ in G. Let $H = \{a^i | i = 0, \pm 1, \pm 2, \ldots\}$; because H is a subgroup of G and $H \neq \{e\}$ (for H contains a, which is not e) our hypothesis on G forces $H = G$.

Let $b = a^2$ and consider $K = \{b^i | i = 0, \pm 1, \pm 2, \ldots\}$, the subgroup generated by b. We have that either $K = \{e\}$ or $K = G$. We consider each of these possibilities.

If $K = \{e\}$ then, since $b \, \varepsilon \, K$, $b = e$, that is to say, $a^2 = e$. But in that case $G = H = \{e, a\}$ and so is a group of order 2.

What happens when $K = G$? In that case, since $a \, \varepsilon \, G = K$, $a = b^j$ for some j. Substituting $b = a^2$, this yields $a = a^{2j}$ from which we deduce that $a^{2j-1} = e$. This tells us that a is of finite order. Let the order of a, $o(a)$, be n. We claim n is a prime number. If not, $n = uv$ where $u > 1$, $v > 1$. Let $c = a^u$; by definition of the order of a, since $u < n$, $c \neq e$. The subgroup generated by c must be all of G. Thus $a = c^k$ leading to $a = a^{uk}$ whence $a^{uk-1} = e$. By Lemma 4.4, $uv = n | (uk - 1)$ which is clearly false. This proves n is a prime number. We leave it to the reader to show that the elements $e, a, a^2, \ldots, a^{n-1}$ are distinct and form a subgroup of G, hence all of G. Thus G has n elements. Because n is a prime our theorem is established.

After we prove Lagrange's theorem (Theorem 4.3 in the next section) we shall see that the converse of this result is also true, namely, a group having a prime number of elements is cyclic and has no nontrivial subgroups (Corollary 1, Theorem 4.3).

Earlier in this section we said that one can exhibit subgroups of a group in two fashions: one, by explicitly taking some specific group and writing down its subgroups, the other, by giving a general construction of classes of subgroups. We already have done this with the subgroups generated by an element in a group. We now proceed to some others.

Let G be a group and let $a \, \varepsilon \, G$. If $b \, \varepsilon \, G$ is such that $ab = ba$, we say that a and b *commute*. The set of all elements in G that commute with a given element a is of some interest. We make the

Definition. If $a \, \varepsilon \, G$, then the *centralizer*, $Z(a)$, of a in G is defined by $Z(a) =$ all x in G with $xa = ax$.

Lemma 4.6. $Z(a)$ is a subgroup of G.

Proof. According to Lemma 4.5 we must verify that:

1. $x, y \, \varepsilon \, Z(a)$ implies $xy \, \varepsilon \, Z(a)$.
2. $x \, \varepsilon \, Z(a)$ implies $x^{-1} \, \varepsilon \, Z(a)$.

Let $x, y \, \varepsilon \, Z(a)$; thus $xa = ax$, $ya = ay$. From this we see that $(xy)a = x(ya) = x(ay) = a(xy)$; in other words, $xy \, \varepsilon \, Z(a)$. Also from $xa = ax$ we have $x^{-1}(xa)x^{-1} = x^{-1}(ax)x^{-1}$. Computing this explicitly we get $x^{-1}(xa)x^{-1} = ax^{-1}$, whereas $x^{-1}(ax)x^{-1} = x^{-1}a$. All in all we have shown that $xa = ax$ implies that $x^{-1}a = ax^{-1}$; in short, $x \, \varepsilon \, Z(a)$ implies $x^{-1} \, \varepsilon \, Z(a)$. We have verified that $Z(a)$ does indeed form a subgroup of G.

We consider some examples of centralizers.

Let $G = S_4$, the symmetric group of degree 4. In G we exhibit the centralizers of some of its elements (see Exercise 4).

EXAMPLE 1.   Let $a = (12)$. What elements in S_4 commute with a? We claim that the four elements e, (12), (34), $(12)(34)$ all commute with a. We leave it to the reader to show that these are all the elements that commute with (12). (Try to find a systematic way of checking this. One hint: If $\sigma\tau$ commutes with a and σ does, then show τ does.)

EXAMPLE 2.   Let $b = (1234) \, \varepsilon \, S_4$. We leave it to the reader to verify that $Z(b) = \{e, b, b^2, b^3\} = \{e, (1234), (13)(24), (1432)\}$.

Another interesting subset of a group, also related to the commuting properties, is the so-called center of the group.

Definition.   The *center* $Z(G)$ of a group G is defined as the set of all a in G satisfying $xa = ax$ for all x in G.

We leave the proof of Lemma 4.7 to the reader.

Lemma 4.7. $Z(G)$ is a subgroup of G.

Exercises*

1. Verify that G and $\{e\}$ are subgroups of G.
2. Verify that in Examples 1, 2, 3, and 5 the subsets cited as subgroups are indeed subgroups.
3. Prove that any group having 4 elements is abelian.
4. (a) Show that in S_4 the only elements commuting with (1234) are $\{e, (1234), (13)(24), (1432)\}$.
   (b) Find all elements in S_4 commuting with (12).

*In all the exercises G denotes a group.

5. Show that the intersection of any two subgroups of G is again a subgroup of G.

6. If H is a subgroup of G and K is a subgroup of H, show that K is a subgroup of G.

7. Show that in any group $Z(e) = G$.

8. Prove that $Z(G)$ is the intersection of all the $Z(a)$ as a ranges over G.

9. Let S be a nonempty set, and let s and t be two different elements of S. Let H be the set of all σ in $A(S)$ leaving s fixed, and K the set of all τ in $A(S)$ leaving t fixed. Prove that there is an element γ in $A(S)$ such that $K = \gamma^{-1}H\gamma$. (Hint: Try defining γ by sending s into t, t into s, and leaving all other elements of S fixed.)

10. Find all the subgroups in a cyclic group of order 24.

11. If G is a cyclic group of order n generated by a, for what values of i will a^i also generate all of G? (Try $n = 6$, 12, and 16, for instance.)

12. Prove that a group having 5 elements must be cyclic (hence abelian) and has no nontrivial subgroups.

13. If H is a subgroup of G and $a \, \varepsilon \, G$, let $a^{-1}Ha$ be the set of all elements $a^{-1}ha$ with h in H. Prove that $a^{-1}Ha$ is a subgroup of G.

14. Call a subgroup H of G *normal* in G if $a^{-1}Ha = H$ for all $a \, \varepsilon \, G$. Prove that the intersection of two normal subgroups of G is a normal subgroup of G.

15. In S_3 find all the normal subgroups and find at least one non-normal subgroup.

4. Lagrange's Theorem

Now that we have seen some examples of subgroups and have obtained some information about them we can try to dig deeper into the interrelationship of a subgroup and the group in which it lies. We want to know how the subgroup sits in the group, and what subsets qualify as possible subgroups. One could ask, for instance, whether a group having 36 elements could have a subgroup having 16 elements. We shall soon see that the answer to this last question is no.

First let us look at an example. Let G be a cyclic group of order 12; in other words G consists of the distinct elements $e, a, a^2, \ldots, a^{11}$ where we assume $a^{12} = e$. Let $H = \{e, a^3, a^6, a^9\}$; a glance reveals that H is a subgroup of G. Consider the subset, which we write as Ha, that we get on multiplying every element in H by a. Writing it out we get $Ha = \{a, a^4, a^7, a^{10}\}$. We note that Ha has four distinct elements; furthermore, Ha has *no element in common* with H. Repeat what we just did for a with a^2; that is, consider $Ha^2 = \{a^2, a^5, a^8, a^{11}\}$. Looking at Ha^2 we see that it, too, has four distinct elements and has no element in common with H or with Ha. If we look at all the elements so obtained (that is, $H \cup Ha \cup Ha^2$) we see that every element of G has been obtained once and only once in H, Ha, or Ha^2. In other words, G has been decomposed as the union of the three mutually disjoint subsets H, Ha, and Ha^2. A decomposition of G as a union of mutually disjoint subsets suggests an equivalence relation on G. With this example in hand and with the motivation it gives us we turn to the general case and introduce an equivalence relation in G based on H.

Let G be a group and H a subgroup of G. We make the

Definition.   If a, $b \; \varepsilon \; G$, then $a \sim b$ if and only if $ab^{-1} \; \varepsilon \; H$. We read this as: *a is congruent to b modulo H.*

Let us examine this in another particular case. Let G be the group of integers under addition; let $n > 1$ be an integer and let H be the set of all multiples of n. H is a subgroup of G. What does the relation introduced in the above definition become in this setting? Because the operation in G is addition, a^{-1} means $-a$, and ab^{-1} really means $a - b$. Thus $a \sim b$ means that $a - b$ is in H; as an element of H, from the form of the elements in H, $a - b$ must be a multiple of n. In other words, here $a \sim b$ means nothing more or less than $a \equiv b$ mod n. We see that the relation introduced in the definition above is an extension, to a much wider context, of the (by now) familiar notion of going modulo n.

We return to the general story. Having defined a relation on G it would be nice if this relation turned out to be decent, an equivalence relation. In fact it does; this is what the next result tells us.

Lemma 4.8.   If G is a group and H a subgroup of G; then the relation "$a \sim b$ if and only if $ab^{-1} \; \varepsilon \; H$" is an equivalence relation on G.

Proof.   We check out in turn the three properties that define an equivalence relation.

1. If $a \; \varepsilon \; G$, then $aa^{-1} = e$; since H is a subgroup of G, e must be in H. Therefore $aa^{-1} \; \varepsilon \; H$ and so $a \sim a$.
2. If a, $b \; \varepsilon \; G$ and $a \sim b$, then $ab^{-1} \; \varepsilon \; H$. Since H is a subgroup of G it contains the inverses of its elements. Thus $(ab^{-1})^{-1} \; \varepsilon \; H$; however $(ab^{-1})^{-1} = (b^{-1})^{-1}a^{-1} = ba^{-1}$. Having $ba^{-1} \; \varepsilon \; H$ we know that $b \sim a$. In other words, $a \sim b$ implies $b \sim a$.
3. If a, b, $c \; \varepsilon \; G$ and $a \sim b$ and $b \sim c$, then $ab^{-1} \; \varepsilon \; H$ and $bc^{-1} \; \varepsilon \; H$. As a subgroup, H contains the product of any two of its elements, hence $(ab^{-1})(bc^{-1}) \; \varepsilon \; H$. But $(ab^{-1})(bc^{-1}) = ac^{-1}$; in short $ac^{-1} \; \varepsilon \; H$ and so $a \sim c$. We have shown that $a \sim b$ and $b \sim c$ forces $a \sim c$.

Having verified that our relation satisfies the three properties that characterize an equivalence relation, we know that it is an equivalence relation.

Whenever one has an equivalence relation one immediately considers the equivalence class of an element. Under our relation "$a \sim b$ if ab^{-1} ε H" what is the equivalence class of any element a in G? Write this class as cl(a). We shall see in a moment that cl(a) is precisely Ha, where this symbol means the set of all elements ha with h running over H. In extended accounts of group theory, where the concept occurs all the time, a name is given to subsets of the kind Ha; they are called *right cosets* of H in G. Of course, there is the analogous notion of left coset (see Exercises 1 and 2).

Note that in the special case where G was the group of integers under addition, and H the subgroup of multiples of n, our previous notation for cl(a) was [a].

We shall prove that cl(a) and Ha are equal by showing that inclusion holds both ways. First we tackle $Ha \subset$ cl(a). If b ε Ha, then $b = ha$ with h ε H, and so $ab^{-1} = h^{-1}$, which is in H since H is a subgroup. Thus b ε cl(a), from the definition of the equivalence relation. Now for the other direction. If d ε cl(a), then $a \sim d$; that is, da^{-1} ε H. This translates into $d = h_1 a$ for some h_1 in H, showing that d ε Ha. Hence cl(a) $\subset Ha$. We record this result as the next lemma.

Lemma 4.9. cl(a) = Ha.

It is natural to ask what the relation must be between the sizes of cl(a) and cl(b) for a, b ε G. Knowing that cl(a) = Ha we expect a certain uniformity in these; why should some element in the group be favored over any other element? The exact relation is given in

Lemma 4.10.   If a, b ε G, then there is a one-to-one correspondence between cl(a) and cl(b), that is, between Ha and Hb.

Proof.   What is the most natural way of setting up this correspondence? After all, every element in Ha is of the form ha where h ε H, and every element in Hb is of the form kb with k ε H. Why not try the mapping ψ: $Ha \rightarrow Hb$ which assigns to ha the element hb? Given x ε Hb then $x = kb$ for some k ε H; hence $x = kb = \psi(ka)$, by the definition of ψ, whence ψ is onto. Is ψ one-to-one? Suppose that $\psi(x_1) = \psi(x_2)$ for x_1, x_2 ε Ha. If $x_1 = h_1 a$ and $x_2 = h_2 a$ with h_1, h_2 ε H, then $\psi(x_1) = h_1 b$, $\psi(x_2) = h_2 b$. Because in G we can cancel, we deduce that $h_1 = h_2$, hence $x_1 = h_1 a = h_2 a = x_2$. In short, $\psi(x_1) = \psi(x_2)$ implies $x_1 = x_2$; that is, ψ is one-to-one. We have shown that ψ is a one-to-one correspondence between Ha and Hb. This proves the lemma.

These preliminary lemmas give us exactly the tools we need to prove an extremely important result due to the French mathematician Lagrange. This result plays a vital role in any discussion of finite groups. It is known as *Lagrange's theorem.*

> J. L. Lagrange (1736–1813) was a great mathematician. He worked in all branches of the mathematics of his time and to each he made fundamental contributions. He left his mark in the theory of equations, number theory, function theory, and mechanics.

Theorem 4.3.   Let G be a finite group of order $o(G)$ and let H be a subgroup of G of order $o(H)$. Then $o(H)$ is a divisor of $o(G)$.

Proof.   We consider the decomposition of G into its equivalence classes under the equivalence relation "$a \sim b$ if $ab^{-1} \varepsilon H$." What do we know about this? By Lemma 4.9, $\mathrm{cl}(a) = Ha$ for all $a \varepsilon G$. By Lemma 4.10 there is a one-to-one correspondence between any Ha and Hb. In the finite case this means that Ha and Hb have the same number of elements. How many elements does Ha have? We consider He; but He is the set of all he with h in H. Hence $He = H$.

Because any Ha has the same number of elements as $He = H$ we have that Ha has $o(H)$ elements. Let k be the number of distinct Ha's. These Ha's do not intersect for they are distinct equivalence classes. Their union is G, for $a \varepsilon \mathrm{cl}(a) = Ha$ for any $a \varepsilon G$. Thus G is the union of the k mutually disjoint Ha's, each of which has $o(H)$ elements. This gives us that $o(G) = ko(H)$. Since k is an integer we see that $o(H)$ is indeed a divisor of $o(G)$. Lagrange's theorem has now been proved.

Definition.   We call k, the number of distinct Ha's, the *index* of H in G.

Lagrange's theorem shows that in a group G of order n, any subgroup must have order dividing n. It is natural to ask whether every possible divisor of n must arise as the order of a subgroup of G. *This in general is not true.* We give an example.

Let G be A_4, the subgroup of all even permutations in S_4. We list the elements of G:

$$G = \{e, (123), (132), (12)(34), (13)(24), (23)(14), (234), (243), (134)$$
$$(143), (124), (142)\}$$

G has no subgroup of order 6 although $6 | 12 = o(G)$. (See Exercise 3.)

We now turn to some implications of Lagrange's theorem. We list these results as corollaries of the theorem.

Corollary 1.   If G is a group with a prime number of elements, then G is cyclic (and hence abelian).

Proof.   Let $a \neq e$ be in G and let H be the subgroup generated by a. If m is the order of H, then m divides the prime $o(G)$. Since $m \neq 1$, we must have $m = o(G)$. But then $H = G$ and so G is indeed cyclic.

Corollary 2.   If G is a finite group and $a \varepsilon G$, then $o(a)$ divides $o(G)$.

Proof.   The definition of $o(a)$, we recall, was the smallest positive integer such that $a^{o(a)} = e$. The powers of a generate a cyclic subgroup H of G. The elements of H are $e, a, a^2, \ldots, a^{o(a)-1}$, and thus H has order $o(a)$. By Lagrange's theorem, $o(H)|o(G)$; because $o(H) = o(a)$ we reach the desired conclusion $o(a)|o(G)$.

Corollary 3.   If G is a finite group of order $o(G)$, then $a^{o(G)} = e$ for all $a \varepsilon G$.

Proof.   Let $a \varepsilon G$ and let $o(a)$ be its order. By Corollary 2 above, $o(a)|o(G)$, and hence $o(G) = ko(a)$. Thus

$$a^{o(G)} = a^{ko(a)} = (a^{o(a)})^k = e^k = e$$

Note that this result sharpens considerably the very crude estimate obtained in the corollary to Lemma 4.3, for we cut down from $o(G)!$ to $o(G)$. Corollary 3 implies, as a very special case, the result we proved earlier in number theory, the so-called "little Fermat theorem" (Theorem 2.8). We do this explicitly.

Corollary 4.   (Fermat). If p is a prime and a an integer such that $p \nmid a$, then $a^{p-1} \equiv 1 \bmod p$.

Proof.   Let $G = \{[1], [2], \ldots, [p-1]\}$ where $[i]$ denotes the congruence class of i modulo p. We have seen that under the multiplication $[i][j] = [ij]$, G forms a group and since $p \nmid a$, $[a] \neq [0]$ so $[a] \varepsilon G$. This group has $p - 1$ elements. Accordingly, by Corollary 3, $[a]^{p-1} = [1]$. Translating this into congruence, because $[1] = [a]^{p-1} = [a^{p-1}]$, we have $a^{p-1} \equiv 1 \bmod p$, as was claimed.

Exercises

1. Let H be a subgroup of G; define, for $a, b \varepsilon G$, $a \mathbin{\#} b$ if $a^{-1}b \varepsilon H$. Prove that $\#$ is an equivalence relation. Note that this is the left analog of the equivalence given in the text.
2. For the relation $\#$ of Exercise 1, prove that $cl(a)$ is the set of all ah with h in H.
3. Show that the group A_4 of order 12 mentioned in the text has no subgroup of order 6.

4. Let G be a cyclic group of order n and let $a \: \varepsilon \: G$ generate G. Describe the order of the subgroup generated by a^i, in terms of i and n. (This exercise is a generalization of Exercise 11 in the preceding section.)

For use in Exercises 5–7 we repeat the definition of the Euler ϕ-function from Exercise 3, Section 6 of Chapter 2. We define $\phi(1) = 1$. For $n > 1$, we define $\phi(n)$ to be the number of positive integers less than n and relatively prime to n.

5. Let $n > 1$ be an integer and consider the set G of classes $[i]$ with i relatively prime to n (as usual, $[i]$ denotes the congruence class of i mod n). Prove that G is a group of order $\phi(n)$ under the product $[i][j] = [ij]$.

6. Using the result of Exercise 5 and Corollary 3 of Lagrange's theorem, prove that if a is an integer relatively prime to n, then $a^{\phi(n)} \equiv 1 \bmod n$.

7. Verify the result of Exercise 6 for $n = 9, 15, 24$, and $a = 7$.

8. Prove that a subgroup of a cyclic group is cyclic.

9. Let G be the symmetric group S_n and H the alternating group A_n $(n > 1)$. Show that H has index two in G, and that the distinct Ha's are just the subsets of even and odd permutations, respectively.

5. Isomorphism

Given a group G, surely the names we assign to its elements should not be important. For instance, consider the group $G = \{e, a\}$ where $a^2 = e$. Suppose we consider the group $G' = \{e', a'\}$ where $(a')^2 = e'$. These groups are not identical, yet we would like to consider them as the same. Let us formalize this idea.

Definition.   A mapping ϕ from a group G to a group G' is said to be an *isomorphism* if:

1. ϕ is a one-to-one mapping of G onto G'.
2. $\phi(ab) = \phi(a)\phi(b)$ for all $a, b \: \varepsilon \: G$.

Note that in 2, in the term $\phi(ab)$, the product ab is computed in G, whereas the term $\phi(a)\phi(b)$ is computed in G' using the product in G'. What 2 says is that ϕ respects (or preserves) the products of elements. The mapping ϕ amounts to a coherent renaming of the elements of G.

We consider some examples.

EXAMPLE 1.   Let G be the group of integers under addition and let G' be the subgroup of even integers. We define $\phi: G \to G'$ by the relation $\phi(x) = 2x$. We have $\phi(x + y) = 2(x + y) = 2x + 2y = \phi(x) + \phi(y)$; and thus ϕ satisfies property 2. Also ϕ is a one-to-one mapping for if $\phi(x) = \phi(y)$, then $2x = 2y$ whence $x = y$. Thus we see that ϕ is an isomorphism of G onto G'.

EXAMPLE 2. Let G be the cyclic group of order n, $G = \{e, a, a^2, \ldots, a^{n-1}\}$ where these are distinct, $a^n = e$, and $a^i a^j = a^{i+j}$, $i + j$ computed modulo n. Let G' be the group Z_n of integers modulo n under addition; $Z_n = \{[0], [1], \ldots, [n - 1]\}$ where $[i] + [j] = [i +j]$, and $[i]$ is the congruence class of i modulo n. Let $\phi: G \to G'$ be defined by $\phi(a^i) = [i]$. We leave it to the reader to prove that ϕ is an isomorphism of G onto G'.

EXAMPLE 3. Let G be the group of all positive real numbers under multiplication and let G' be the group of all real numbers under addition. Define $\phi: G \to G'$ by $\phi(x) = \log x$ for $x \, \varepsilon \, G$. Note that to verify $\phi(xy) = \phi(x)\phi(y)$, since the multiplication on the right-hand side is that of G', and since the operation in G' is addition, we must show that $\phi(xy) = \phi(x) + \phi(y)$. Using $\phi(x) = \log x$, we see that this relation translates into $\log (xy) = \log x + \log y$; this is true, being one of the basic properties of the logarithm function. Thus ϕ satisfies the requisite property 2 in the definition of an isomorphism. That ϕ moreover is one-to-one and onto is another basic property of the logarithm.

EXAMPLE 4. We reverse the roles of the two groups in Example 3. Let G be the group of real numbers under addition, and let G' be the group of positive real numbers under multiplication. Define $\psi: G \to G'$ by $\psi(x) = 10^x$ for $x \, \varepsilon \, G$. By well-known properties of the function 10^x, ψ is indeed one-to-one and onto. We claim that $\psi(x + y) = \psi(x)\psi(y)$—this is what we must verify in order to establish property 2, because the operation in G is addition. To see this, we merely note from the definition of ψ that $\psi(x + y) = 10^{x+y} = 10^x 10^y = \psi(x)\psi(y)$. Thus ψ is an isomorphism of G onto G'.

Note that the functions ϕ and ψ of Examples 3 and 4 are inverse functions; that is, $\psi = \phi^{-1}$. It is easy to prove that if θ is an isomorphism of G onto G', then θ^{-1} is an isomorphism of G' onto G (see part (a) of Exercise 1).

It is also not difficult to prove that if θ is an isomorphism of G onto G' and ψ is an isomorphism of G' onto G'', then the product of θ and ψ is an isomorphism of G onto G'' (see part (b) of Exercise 1).

From the properties of isomorphisms just mentioned it follows that the relation of isomorphism between groups is an equivalence relation.

When there is an isomorphism between two groups, we say that these groups are isomorphic.

We consider the action of an isomorphism on the identity element and on inverses.

Lemma 4.11. If ϕ is an isomorphism of G onto G' then:

    1. $\phi(e) = e'$, the identity element of G'.
    2. For any $a \varepsilon G$, $\phi(a^{-1}) = \phi(a)^{-1}$.

Proof. For x in G we have $x = xe$ so that $\phi(x) = \phi(xe) = \phi(x)\phi(e)$. Since $\phi(x) = \phi(x)e'$, we get $\phi(x)\phi(e) = \phi(x)e'$; but since we can cancel in a group we obtain $\phi(e) = e'$.

If $a \varepsilon G$, then $aa^{-1} = e$, so that $\phi(aa^{-1}) = \phi(e)$, which we have just shown to be equal to e'. Moreover $\phi(aa^{-1}) = \phi(a)\phi(a^{-1})$. The upshot of this is that $\phi(a)\phi(a^{-1}) = e'$. This tells us that $\phi(a^{-1}) = \phi(a)^{-1}$. The lemma is now proved.

If one examines the proof just given one sees that neither the "one-to-one-ness" nor the "onto-ness" of ϕ played any role; what entered crucially was the fact that $\phi(ab) = \phi(a)\phi(b)$. Such mappings are called *homomorphisms*. In other words a mapping $\psi: G \to G'$ is a homomorphism if $\psi(ab) = \psi(a)\psi(b)$ for all a and b in G.

We have encountered homomorphisms before. One place they appeared was in Theorem 3.7; there it was shown that the signature $\Lambda(\sigma)$ of a permutation σ furnishes a homomorphism of the symmetric group onto the multiplicative group $\{1, -1\}$.

For another instance of a homomorphism let G be the additive group of integers and G' the additive group of integers modulo n. Then the mapping that assigns to an integer $i \varepsilon G$ its congruence class $[i]$ modulo n is a homomorphism of G onto G'. Some examples and properties of homomorphisms appear in the exercises.

We conclude this chapter with an old theorem of Cayley which asserts that every group can be realized as a subgroup of the group of one-to-one mappings of some set onto itself. In particular, if the group is finite, the theorem asserts that the group must be isomorphic to some subgroup of S_n for an appropriate integer n. Because our concern will be with permutations, we retain here and in the exercises the product of functions that was used in Chapter 3. In other words, *if f and g are one-to-one mappings of a set S onto itself, then by fg we mean the mapping of S onto itself defined by $(fg)(x) = g(f(x))$.*

We proceed to

Theorem 4.4. (Cayley). Let G be a group. Then G is isomorphic to a subgroup of $A(S)$ for some set S.

Arthur Cayley (1821–1895) was an English mathematician who worked primarily in what could be described as the algebra of his time. He was a leading expert on invariant theory. In the 1854 paper from which Theorem 4.4 is taken, he came very close to the modern definition of a group.

Proof. Recall that $A(S)$ denotes the group of all one-to-one mappings of S onto itself.

The theorem says that for some set S, $A(S)$ will be suitably related to G. Where can we possibly dig up this S? The most natural place is G itself.

Consider $G = S$ merely as a set. If $g \ \varepsilon\ G$, let $T_g : G \to G$ be defined by $T_g(x) = xg$ for all x in G. We assert that $T_g \ \varepsilon\ A(S)$. Given $x \ \varepsilon\ G$ then $x = (xg^{-1})g$, which is to say, $x = T_g(xg^{-1})$. Therefore T_g is onto. Is it one-to-one? Suppose that $T_g(x) = T_g(y)$; this translates into $xg = yg$. Because we are in a group we may cancel the g; this yields $x = y$. In short, we have shown that T_g is a one-to-one mapping of $S(= G)$ onto itself. In other words, $T_g \ \varepsilon\ A(S)$.

We compute $T_g T_h$ for $g, h \ \varepsilon\ G$. Now

$$(T_g T_h)(x) = T_h(T_g(x)) = T_h(xg) = (xg)h = x(gh) = T_{gh}(x)$$

This tells us that $T_g T_h = T_{gh}$. What is T_g^{-1}? Now

$$(T_{g^{-1}} T_g)(x) = T_{g^{-1}g}(x) = T_e(x) = xe = x$$

Therefore $T_{g^{-1}} = T_g^{-1}$.

Let G' be the set of all T_g in $A(S)$ for $g \ \varepsilon\ G$. We claim that G' is a subgroup of $A(S)$. Having shown that $T_g T_h = T_{gh}$, we have that G' is closed under the product of $A(S)$. Since $T_g^{-1} = T_{g^{-1}}$, we have that G' possesses the inverses of its elements. From this, G' is a subgroup of $A(S)$—see Lemma 4.5. G' will turn out to be the subgroup required in the assertion of the theorem.

Let $\psi : G \to G'$ be defined by $\psi(g) = T_g$. From the definition of G', ψ is onto. Is ψ one-to-one? If $\psi(g) = \psi(h)$ for $g, h \ \varepsilon\ G$, then $T_g = T_h$; hence $T_g(x) = T_h(x)$ for all x in G. In particular this is true with $x = e$, and so we get $eg = eh$ and $g = h$. Therefore ψ is a one-to-one mapping of G onto G'.

Consider $\psi(gh)$ for g and h in G. Now $\psi(gh) = T_{gh}$. We have seen that $T_{gh} = T_g T_h$. Thus $\psi(gh) = T_{gh} = T_g T_h = \psi(g)\psi(h)$. We have shown that ψ is an isomorphism of G onto G'. With this the theorem is proved.

We specialize to the finite case. In that situation it becomes the

Corollary. A finite group is isomorphic to some group of permutations of a finite set, that is, to a subgroup of S_n for some n.

We exhibit how to calculate this permutation representation for two particular finite groups.

1. Let G be a cyclic group of order n, $G = \{e, a, a^2, \ldots, a^{n-1}\}$. What is the mapping T_a on G? Remember that $T_a(x) = xa$ for $x \varepsilon G$. We number the group elements as follows:

$$x_1 = e, x_2 = a, \ldots, x_i = a^{i-1}, \ldots, x_n = a^{n-1}$$

Now

$$T_a(x_1) = x_1 a = ea = a = x_2, T_a(x_2) = x_2 a = aa = a^2 = x_3, \ldots,$$
$$T_a(x_{n-1}) = x_{n-1}a = a^{n-2}a = a^{n-1} = x_n$$

What is $T_a(x_n)$? Now

$$T_a(x_n) = x_n a = a^{n-1}a = a^n = e = x_1$$

Thus T_a, as a permutation on the x's, is nothing but the cycle $(12 \cdots n)$. The permutation T_{a^i}, representing a^i, is $(12 \cdots n)^i$, since we know that $T_{a^i} = (T_a)^i$.

2. Let G be the group $\{e, a, b, ab\}$ where $a^2 = b^2 = e$ and $ab = ba$. We number the elements as $x_1 = e$, $x_2 = a$, $x_3 = b$, and $x_4 = ab$. What is T_a?

$$T_a(x_1) = x_1 a = ea = a = x_2, T_a(x_2) = x_2 a = aa = a^2 = e = x_1,$$
$$T_a(x_3) = x_3 a = ba = ab = x_4, \text{ and } T_a(x_4) = x_4 a = (ab)a$$
$$= (ba)a = ba^2 = be = b = x_3$$

Therefore

$$T_a = \begin{pmatrix} 1\ 2\ 3\ 4 \\ 2\ 1\ 4\ 3 \end{pmatrix} = (12)(34)$$

One can compute and get $T_b = (13)(24)$ and $T_{ab} = (14)(23)$. Note that

$$T_{ab} = (14)(23) = (12)(34) \cdot (13)(24) = T_a T_b$$

as it should be.

Exercises

1. (a) If ϕ is an isomorphism of G onto G', prove that ϕ^{-1} is an isomorphism of G' onto G.
   (b) If θ is an isomorphism of G onto G' and ψ is an isomorphism of G' onto G'', prove that the product of θ and ψ is an isomorphism of G onto G''.
2. Prove that a cyclic group of order n is isomorphic to Z_n.
3. Prove that any two cyclic groups of order n are isomorphic.

4. Prove that the mapping $\phi: G \to G'$ defined by $\phi(a) = e'$ for all a in G is a homomorphism. (It is called the *trivial* homomorphism.)

5. If ϕ is a homomorphism of G into G' let $K(\phi)$ be the set of all x in G with $\phi(x) = e'$. ($K(\phi)$ is called the *kernel* of ϕ.) Prove:
   (a) $K(\phi)$ is a subgroup of G.
   (b) If $a \varepsilon K(\phi)$ and $x \varepsilon G$, then $x^{-1}ax \varepsilon K(\phi)$. (Thus $K(\phi)$ is a normal subgroup of G; see Exercise 14 in Section 3.)

6. Let G be the group of integers under addition and let $G' = \{e, a\}$ be a cyclic group of order two. Define $\phi: G \to G'$ by setting $\phi(n) = a$ if n is odd and $\phi(n) = e$ if n is even. Prove that ϕ is a homormorphism of G onto G'. Identify $K(\phi)$.

7. Let G be the group S_3. Compute T_a (see Theorem 4.4) for all a in G. (This represents S_3 as a group of permutations on six objects.)

8. Let G be a finite abelian group and let $n > 1$ be an integer. Define $\phi: G \to G$ by $\phi(a) = a^n$ for all a in G.
   (a) Show that ϕ is a homomorphism.
   (b) What is the kernel $K(\phi)$?

9. If there exists a homomorphism of G *onto* G' and G is abelian, prove that G' must also be abelian.

10. Give an example showing that the converse of Exercise 9 is false; that is, if G' is abelian it need not follow that G is abelian.

11. Let ϕ be a homomorphism of G into G' and let $a \varepsilon G$ be of finite order. Prove that $o(\phi(a))|o(a)$. If ϕ is an isomorphism, show that the two are equal.

12. Let G be a cyclic group and suppose that there exists a homomorphism of G onto G'. Prove that G' is also cyclic.

13. Let G be any group and let $a \varepsilon G$. Define $\phi_a: G \to G$ by $\phi_a(x) = a^{-1}xa$ for any $x \varepsilon G$.
   (a) Prove that ϕ_a is an isomorphism of G onto itself.
   (b) Characterize those $a \varepsilon G$ for which ϕ_a is the identity mapping of G onto itself.
   (c) If $a, b \varepsilon G$, prove that $\phi_{ab} = \phi_a\phi_b$.

14. Let G be a group and let $A(G)$ be the group of one-to-one mappings of G onto itself. Define $\psi: G \to A(G)$ by $\psi(a) = \phi_a$ (see Exercise 13). Prove that ψ is a homomorphism of G into $A(G)$. Identify $K(\psi)$.

15. If ψ is a homomorphism of G onto G' with kernel $K(\psi)$, prove that ψ is an isomorphism if and only if $K(\psi) = \{e\}$.

16. Let ψ be a homomorphism of G onto G' with kernel $K(\psi)$. Let g' be an element of G' and suppose that $g' = \psi(g)$ with g in G. Show that every element x in G such that $\psi(x) = g'$ must be of the form $x = gk$ with $k \varepsilon K(\psi)$.

17. Call an isomorphism of G onto itself an *automorphism* of G. Prove that the set of automorphisms of G forms a subgroup of the group of all one-to-one mappings of G onto itself.
18. Find all automorphisms of the additive group of integers.
19. Find all automorphisms of a cyclic group of order n.
20. Let G be the group $\{e, a, b, ab\}$ where $a^2 = b^2 = e$ and $ab = ba$. Find all automorphisms of G.
21. Let G be any group. For a in G define $L_a: G \to G$ by $L_a(x) = a^{-1}x$. Prove that the set of all L_a's forms a subgroup G' of $A(G)$. Prove that G' is isomorphic to G.
22. Let G be a finite abelian group of order n, with elements $a_1, \ldots, a_n$. Let $x = a_1 a_2 \cdots a_n$. Prove:
    (a) $x^2 = e$
    (b) If G has exactly one element $a \neq e$ with $a^2 = e$, then $x = a$.
    (c) If G has more than one element $a \neq e$ with $a^2 = e$, then $x = e$. (Warning: Part (c) is not easy.)
23. Use the result of part (b) of the preceding exercise to prove Wilson's theorem (Theorem 2.9).

Chapter 5

Finite Geometry

1. Introduction

From sets, number theory, and groups, we turn in this chapter to a fourth fundamental subject—geometry.

Geometry has a claim to being mankind's oldest mathematics, although some simple arithmetic may have preceded the earliest geometrical discoveries. But there is no doubt that geometry, as developed by the Greeks, was the first substantial mathematics (or at least, the first of which we have any knowledge).

About 300 B.C. a man named Euclid (about whom we know very little else) wrote a book called the *Elements*. Other works by Euclid have survived, but he is famous for the *Elements*. In this masterpiece he organized, arranged, and perfected most of the mathematics known up to his time. It is by far the most successful textbook ever written. It is in wide use today in virtually the same form that he wrote it. In moderately altered form, it is the basis of the geometry taught to the majority of today's high-school students.

We shall not study geometry in the direct tradition of Euclid. Our project will be a little different, and in particular it will lead to geometries that would surely have intrigued Euclid. To gain historical perspective, some comparison with Euclid will be enlightening. Let us therefore open his remarkable book and read a little. We happen to be taking the quotations from the edition published in 1933 by J. M. Dent in a collection called *Everyman's Library*; the editor was Heath, and he used earlier versions edited by Simson and Todhunter.

Euclid starts right off with 35 definitions. Here are the first four.

1. A point is that which has no parts, or which has no magnitude.
2. A line is length without breadth.
3. The extremities of a line are points.
4. A straight line is that which lies evenly between its extreme points.

Styles of mathematical exposition have changed. These are highly descriptive phrases, somewhat in the manner of dictionary definitions. They are not mathematical definitions in the sense that became standard toward the end of the nineteenth century. Our introduction of points and lines in the next section will present an interesting contrast.

Following the 35 definitions we find 3 postulates and 12 axioms. The first axiom reads:

AXIOM 1.   Things which are equal to the same thing are equal to each other.

As was appropriate for a treatise that aimed to cover nearly all the mathematics of its day, Euclid's *Elements* thus paid attention to logical as well as to geometrical foundations. There are a number of further axioms in the same vein. The next to the last axiom does have a striking geometric content:

AXIOM 11.   All right angles are equal to one another.

But nothing up to this point has prepared us for the astonishing final axiom.

AXIOM 12.   If a straight line meets two straight lines, so as to make the two interior angles on the same side of it taken together less than two right angles, these straight lines, being continually produced, shall at length meet on that side on which are the angles which are less than two right angles.

The axiom is illustrated in Figure 5.1; the sum of the marked angles is less than 180 degrees, and the conclusion is that the lines meet when extended to the right.

Some dictionaries define an axiom to be a "self-evident truth." Through the centuries, many people questioned the self-evident nature of Euclid's Axiom 12, and eventually began to question its truth. One thing

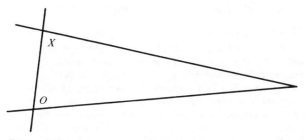

Figure 5.1

these inquiries accomplished at an early stage was the discovery of alternate versions of the axiom, which might be considered more palatable. The most popular of these was called the "parallel postulate."

PARALLEL POSTULATE.   Given a line *L* and a point *P* not on *L*, there exists exactly one line through *P* parallel to *L*.

The parallel postulate will play a key role in our development of geometry.

To conclude this introductory glimpse of Euclid's geometry, let us look for a moment at how he continues, after the definitions, postulates, and axioms have been set down. The initial heading is "Book I, Proposition 1, Problem." The problem in question (today it is more suggestive to call it an existence theorem) reads: to describe an equilateral triangle on a given finite straight line. If the proposed base is *AB*, Euclid's construction is to draw the circles with centers *A* and *B* and radius *AB*. If the two circles intersect, our problem is obviously solved. For over 2000 years no one questioned the obvious assertion that the circles intersect. Today our point of view is that we must add some axioms to Euclid's list to make certain that the circles intersect.

When such a modern axiom system is set forth, we find ourselves able to assess the role of the parallel postulate. On omitting it from the axioms, exactly three possibilities emerge. (1) We can affirm the parallel postulate, the result being Euclidean geometry. (2) We can deny it by asserting that there are many parallels; this leads to hyperbolic geometry, discovered independently by Bolyai and Lobachevski. (3) We can deny it by asserting that there are no parallels; then we get elliptic geometry, discovered by Riemann.

Janos Bolyai (1802–1860), a Hungarian mathematician, developed hyperbolic geometry in collaboration with his father Farkas Bolyai, also a mathematician. There are indications that Gauss knew the theory earlier, but refrained from publication because he thought it might stir up controversy. Independently of Bolyai, and at almost the same time, Nikolai Ivanovich Lobachevski (1793–1856) published the discovery in Russia. Many nonmathematicians first heard of Lobachevski in an amusing song recorded by Tom Lehrer.

Thomas Andrew Lehrer (1928–) became a noted song writer after a mathematical career at Harvard that included two collisions with one of the authors of this book. Some (but not all) of his songs are available in three hilarious records, or you can try one of his songbooks. His recent activities have included the teaching of political science and mathematics for social scientists at M.I.T. and the University of California at Santa Cruz.

George Friedrich Bernhard Riemann (1826–1866) took his Ph.D. at Göttingen in 1854 under Gauss's direction. Earlier, he had served in the Prussian army under Frederick William IV, helping to stamp out the revolution of 1848. His brief career ended in his death from tuberculosis at Lago Maggiore, Italy. Mathematics owes to him a fabulous number of highly original ideas. Most mathematicians consider the Riemann hypothesis (a certain statement about the zeros of the Riemann zeta function) the most important unsolved problem in mathematics.

2. Affine Planes

In beginning our study of geometry, the first task is to decide how we shall introduce points and lines. The framework of set theory, as introduced in Chapter 1, serves the purpose well. "Points" are to be elements of a certain set. Our notation for the set will be Π, the capital form of the Greek letter pi. (Π stands for "plane"; we do not use P, reserving that letter for points.) "Lines" are to be certain subsets of Π. The letters L, M, N will typically be used for lines.

Before proceeding we make a remark on terminology. Instead of saying that a point P is a member of a line L, we shall normally use geometric language, and say that "P lies on L" or that "L goes through P." We further say that points are *collinear* if they lie on a line.

Naturally we have to introduce some restrictions before this can be interesting or useful. If no axioms are added we might, for instance, take as lines all the subsets of Π (as soon as Π has at least three points, this will be ruled out by Axiom 1 below.) At the other extreme we might assume that there are no lines at all (as soon as Π has at least two points, this too is ruled out by Axiom 1). In between, absolutely any selection of subsets of Π, however weird, is conceivable until we put down some axioms.

We present our first axiom.

AXIOM 1. Given any two distinct points P and Q of Π, there is exactly one line going through P and Q.

Let us experiment a little with the implications of Axiom 1. Let us suppose that the points of Π are given by numbers 1, 2, . . . , and let us use the notation {. . .} of Chapter 1 for subsets. For any i and j there must be a line (i.e., special subset) of the form $\{i, j, . . .\}$, and exactly one, through them. What are the possibilities?

A first remark is that there is nothing in Axiom 1 to exclude lines with exactly one point, or even having the null set as a line (very soon Axiom 3 will rule this out). Such lines would play no role in fulfilling (or for that matter violating) Axiom 1. So let us simplify the present discussion by assuming that all lines have at least two points.

Perhaps the simplest way to fulfill Axiom 1 is to take the lines to be *all* two-point subsets of Π. (See Exercise 3.)

As a second example, suppose the points are 1, 2, 3, 4. Suppose one of the lines is {1, 2, 3}. Then no other line is allowed to contain two of 1, 2, 3. There must be a line containing 1 and 4, and it must be {1, 4} itself. Similarly {2, 4} and {3, 4} are lines. There can be no other lines. We recapitulate the list, and note that Axiom 1 is fulfilled:

$$\{1, 2, 3\}, \quad \{1, 4\}, \quad \{2, 4\}, \quad \{3, 4\}$$

Further examples of this kind occur in Exercises 1 and 2.

Let L and M be two distinct lines of Π. What can the intersection $L \cap M$ be? It cannot contain as many as two points, for this would violate Axiom 1. So the possibilities are that L and M meet in exactly one point, or that they do not meet at all.

Definition. Two lines are *parallel* if they do not intersect.

Our second axiom will be the parallel postulate that was discussed in the preceding section.

AXIOM 2. Given any line L and a point P not on L, there is exactly one line through P parallel to L.

We could begin the theory on the basis of Axioms 1 and 2. But it turns out that a number of dull extreme cases would be included (these are treated in Exercises 5–7), and the sensible thing to do is to throw in extra axioms with the deliberate purpose of excluding these uninteresting cases. This can be done in several different ways. Our choice is to add the following two axioms.

AXIOM 3. Every line contains at least two points.

AXIOM 4. There exist at least two lines.

It is time to give a name to the object we are discussing. The naked term *plane* might be a little misleading. In any event, there is a standard name in the mathematical literature, and we shall adopt it; the name in question is *affine plane*. (It would be fascinating but too much of a digression to trace the origin of the word *affine*, and explain the way affine planes contrast with other planes that today's geometers study.)

We summarize in a definition.

Definition. An *affine plane* has the following elements of structure:

1. a set Π, whose members we call "points"
2. certain special subsets of Π, which we call "lines"
3. the assumption that Axioms 1, 2, 3, and 4 hold

Our assumptions are so modest that it may seem unlikely that a good deal of interesting geometry can now be developed. Nevertheless, this is the case. In a sense, nearly all geometry can be reached from this starting point. The ordinary plane of Euclid's geometry of course satisfies our axioms, and so everything we do ought to apply to it. But note that in the Euclidean plane every line of course contains an infinite number of points. There is nothing in our axioms to preclude the possibility that lines might have only a finite number of points. Indeed this is just the case that our discussion will emphasize. Naturally, when we assume explicitly that lines have only a finite number of points, we depart from Euclidean geometry. But Euclid's plane still serves, in good part, as a model to guide us as we develop novel finite geometry.

Exercises

1. Let $\Pi = \{1, 2, \ldots, n\}$ with $n \geq 3$. Suppose one of the lines is $\{1, 2, \ldots, n - 1\}$, and that Axiom 1 is satisfied.
   (a) Prove that the subsets $\{1, i\}$, $i = 2, \ldots, n$, must also be lines.
   (b) Prove that there are no other lines.
   (c) Which of Axioms 2, 3, and 4 are satisfied?
2. Let $\Pi = \{1, 2, 3, 4, 5\}$. Suppose one of the lines is $\{1, 2, 3\}$ and that Axiom 1 is satisfied.
   (a) If $\{4, 5\}$ is a line, prove that the remaining lines are $\{1, 4\}$, $\{1, 5\}$, $\{2, 4\}$, $\{2, 5\}$, $\{3, 4\}$, and $\{3, 5\}$.
   (b) If $\{4, 5\}$ is not a line, prove that exactly one of $\{1, 4, 5\}$, $\{2, 4, 5\}$, and $\{3, 4, 5\}$ must be a line.
   (c) If, for instance, $\{1, 4, 5\}$ is a line, prove that the remaining lines are $\{2, 4\}$, $\{2, 5\}$, $\{3, 4\}$, and $\{3, 5\}$.
   (d) In each case determine which of Axioms 2, 3, and 4 are fulfilled.
3. Let Π be a set of n elements. Let the lines be all two-element subsets of Π. Prove that this makes Π an affine plane for $n = 4$, but fails to do so for any other n.
4. Let Π be a set of n elements, $n \geq 3$. Let the lines be all two-element subsets, and in addition, the whole set Π.
   (a) Prove that Axioms 3 and 4 are fulfilled, and that Axiom 1 is not.

(b) Prove that Axiom 2 is fulfilled for $n = 4$, but for no other value of n.

The following set of exercises is devoted to determining the possibilities when Axioms 1 and 2 are fulfilled, but Axiom 3 or Axiom 4 fails. The case where Π is the null set is a little ridiculous, and we brush it aside. Exercise 5 disposes of the case (almost equally ridiculous) where Π has one point, after which we can assume at least two points in Π.

5. Let Π be a set containing just one point P. Show that the following are the ways lines can be chosen so as to fulfill Axioms 1 and 2.
   (a) no lines at all
   (b) just $\Pi = \{P\}$
   (c) Π and the null set
6. Let Π be any set. Show that in the following cases the lines satisfy Axioms 1 and 2.
   (a) The only line is Π.
   (b) The lines are Π and the null set.
   (c) The lines are Π and all one-element subsets.
7. Let Π be a set with at least two elements. Suppose that we are given a collection of lines (i.e., subsets of Π) satisfying Axioms 1 and 2 but violating Axiom 3 or Axiom 4, or both. Prove that the lines must be as in (a), (b), or (c) of Exercise 6.
8. Let Π be the set of all points in the interior of a circle C. Let the lines of Π be all "open line segments" inside C: that is, for any two distinct points x and y inside C, take the segment joining x and y, not including the endpoints x, y. Which of the axioms are fulfilled?
9. Let Π be the set of points on the surface of a sphere, and let all the lines be great circles. Which of the axioms are fulfilled?
10. Take for Π all the points of the ordinary Euclidean plane, except that the origin 0 is deleted. Take for "lines" all the ordinary lines through 0 and also all circles through 0 (in each case, of course, the point 0 is missing from the "line"). Which of the axioms are fulfilled?
11. Someone is trying to make a plane out of the points $\{1, 2, 3, 4, 5, 6\}$. He has decided to take $\{1, 2, 3\}$ and $\{4, 5, 6\}$ as lines and will presumably add additional lines. He is requiring that all lines have at least two points and that through any two points there should be exactly one line.
    (a) Complete his list of lines.
    (b) Does he get an affine plane? (In other words, does the parallel postulate hold?)

3. Counting Arguments

Our first objective is to prove that any two lines in an affine plane have the same number of points. This is not the place to discuss the question of what this means if one or both lines contain an infinite number of points, so we now make the blanket assumption: *In the remainder of this chapter (except for the example of the Moulton plane in Section 6) all affine planes are finite, that is, they contain only a finite number of points.* To avoid tedious repetition, finiteness will not be explicitly mentioned in the theorems; it is to be tacitly assumed in all theorems from Theorem 5.1 on.

After infinite numbers have been introduced in Chapter 7, we shall be able to say unreservedly that any two lines in an affine plane have the same number of points.

The proof of Theorem 5.1 will be a nice illustration of the ideas in Chapter 1. We are given two lines, say *L* and *M*, and we have to prove that *L* contains the same number of points as *M*. One might imagine counting the points on *L*; in a particular example, maybe we would find 17. Then we might separately count the points on *M*, and (if all is well) find 17 again. But instead of such a double counting operation, we are going to *construct a mapping F* from *L* to *M* and prove that *F* is one-to-one and onto. Of course this mapping proves that *L* and *M* contain the same number of points, and, in a way, it does so without finding out what that number is.

The method is particularly easy to visualize for the case of two parallel lines (see Figure 5.2). Suppose we are given two parallel lines *L* and *M*, and a point *P* not on either *L* or *M*. A typical line through *P* meets *L* in *A* and *M* in *B*. We make *A* correspond to *B*, and we can argue that this sets up a one-to-one correspondence between the points of *L* and those of *M*.

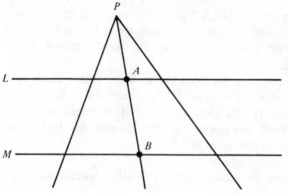

Figure 5.2

However, we shall not use this argument. For one thing, the point P not on L or M does not always exist (see Exercise 1). More important, we would still have to do the case of two intersecting lines, and it will turn out that the case of parallel lines is very easily reduced to the case of intersecting lines. (The proof of Theorem 5.1 will in fact begin with this reduction.)

Theorem 5.1. Any two lines in an affine plane have the same number of points.

Proof.   We first make the promised reduction to the case of intersecting lines.

Let then L and M be two parallel lines. Pick any point A on L and any point B on M (Figure 5.3). Consider line AB (we use the notation of elementary geometry for the line joining two points; note that line AB usually contains many points in addition to A and B). It intersects both L and M. If we have proved Theorem 5.1 for the case of intersecting lines, then we know that AB and L have the same number of points that AB and M do. Hence we can conclude that L and M have the same number of points.

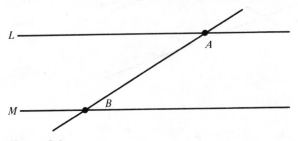

Figure 5.3

So it remains for us to handle the case where L and M intersect, say in point C (Figure 5.4). This time we need an auxiliary point P not lying on either L or M. So as not to interrupt the main line of the argument, we leave to the end of the proof (as a lemma) the argument that such a point P exists.

In outline the proof is this: Take a typical line through P. It will meet L in A and M in B. We make A and B correspond. There is one catch to this. Suppose the line we take through P is parallel to M (Figure 5.5). Then it will meet L in a point X, and we have no point on M to pair with it. Similarly, the line through P parallel to L (Figure 5.6) meets M in an orphaned point Y. Faced with this double difficulty, we can restore order by making X and Y correspond.

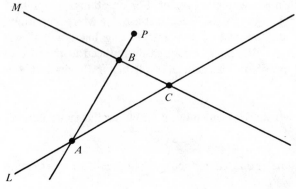

Figure 5.4

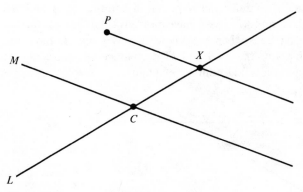

Figure 5.5

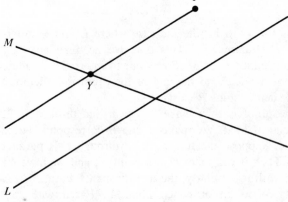

Figure 5.6

We proceed to give the full details. We shall carefully define a mapping from the points of L to those of M, and then prove that the mapping is one-to-one and onto.

We begin by discussing points X and Y. Note that we chose P so that P does not lie on L or M. By Axiom 2, there is a unique line through P parallel to M; call it M_0. Because M_0 contains P which is not on L, M_0 is certainly not the same line as L. Can M_0 be parallel to L? If so, both M and L are parallel to M_0 and so (see Exercise 2) M and L are either parallel or identical. This being impossible (M and L meet precisely at point C), we conclude that M_0 meets L at just one point, point X. In the same way the line through P parallel to L meets M at point Y.

We proceed to define the desired mapping F from L to M. Take any point A on L. Case I: If A is the point X, we set $F(X) = Y$. Case II: A is not X. We consider line AP. It is certainly not identical to M, since it contains P, which is not on M. Can AP be parallel to M? If so, AP is the *unique* line through P parallel to M, and we agreed that it meets L in X, whereas $A \neq X$. Hence AP meets M at point B, and we take $B = F(A)$. (Note: The reader might worry a little that point C needs special treatment, but there is actually nothing extra that needs to be said about C. Of course, $F(C) = C$.)

We now have a map F defined on all of L and going to M. We have a choice of two methods for proving that F is one-to-one and onto: We can try to hit it head on, or we can invent a suitable function G and prove that the products FG and GF are both equal to the identity (see Exercise 33 in Chapter 1, Section 3). We leave the second option as Exercise 3, and proceed to a frontal assault on F.

F IS ONE-TO-ONE. Suppose that $F(A_1) = F(A_2)$; we have to prove that $A_1 = A_2$. We distinguish two cases. (1) A_1 and A_2 are different from X. In this case both $F(A_1)$ and $F(A_2)$ are obtained by joining A_1 or A_2 to P and finding where the line meets M. If $F(A_1) = F(A_2) = B$, then both A_1 and A_2 are on the line PB, so $A_1 = A_2$ equals the point where PB meets L. (2) One of the points A_1, A_2 is X. Say $A_2 = X$. By definition $F(X) = Y$. If $A_1 \neq X$ then $F(A_1)$ is obtained by intersecting PA_1 with M. So $F(A_1) = Y$ means PA_1 meets M at Y. But PY is the line through P parallel to L (see Figure 5.6) and cannot contain A_1. Hence $A_1 = X$ is the only tenable conclusion.

F IS ONTO. Take a point B on M. We need to find a point A on L with $F(A) = B$. Again there are two cases. If $B = Y$, then $A = X$ will do. If $B \neq Y$, then the line PB is not parallel to L. Hence it meets L, and the point of intersection A is the point we want.

To conclude the proof of Theorem 5.1 we still have to provide the proof promised of the following lemma.

Lemma.   Let L and M be any intersecting lines in an affine plane, $L \neq M$. Then there exists a point P not on L or M.

Proof.   Say L and M intersect in C (Figure 5.7). By Axiom 3 there is at least one point in addition to C on each of L and M. Pick A on L and B on M. Line AB cannot contain C (for then AB would be the same line as L and M). Draw through C the unique line parallel to AB (Axiom 2), and pick P on it, $P \neq C$ (Axiom 3). Then P cannot lie on L for PC would be line L and would meet AB at A, instead of being parallel to AB. Similarly, P is not on M.

We have thus concluded the proof of Theorem 5.1. Now that we know that all lines in an affine plane have the same number of points, it is natural to give a name to this common number; we call it the *order* of Π, usually denoted by n.

Theorem 5.2.   Let Π be an affine plane of order n. Let L be a line in Π. Then Π contains exactly $n - 1$ lines parallel to L.

Proof.   Pick any line M distinct from L and intersecting L. (To assure ourselves that such a line exists we have only to pick a point A on L, a point B not on L, and let $M = AB$.) Line M contains exactly n points. One of these is its point of intersection with L; let us call this A_1, and write $A_2, \ldots, A_n$ for the remaining points on M (Figure 5.8).
     By Axiom 2 there is exactly one line through A_2 parallel to L;

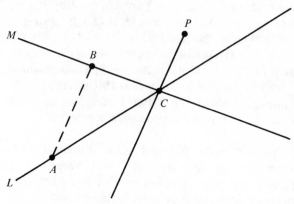

Figure 5.7

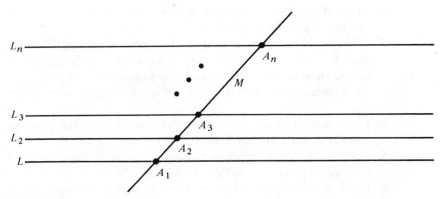

Figure 5.8

call it L_2. Similarly we have $L_3, \ldots, L_n$ through $A_3, \ldots, A_n$ parallel to L. This gives us $n-1$ lines parallel to L. So let L_o be parallel to L. Then (see Exercise 2) L_o cannot be parallel to M. Lines L_o and M must meet in one of the points $A_1, A_2, \ldots, A_n$ of M. That point cannot be A_1, for A_1 lies on L and L_o is parallel to L. Hence L_o must meet M in one of the points $A_2, A_3, \ldots, A_n$. Suppose, for instance, that L_o and L meet in A_3. Then L_o and L_3 are both lines through A_3, parallel to L. By the uniqueness part of Axiom 2, $L_o = L_3$. Thus the general line L_o parallel to L must be one of $L_2, L_3, \ldots, L_n$.

Theorem 5.3. Let Π be an affine plane of order n. Then the total number of points in Π is n^2.

Proof.   We use Theorem 5.2, and maintain its notation. With L any line in Π, we have $L_2, L_3, \ldots, L_n$ as the complete list of lines parallel to L. Counting L as well, we have a total of n lines, each containing exactly n points. There is no overlapping between any two of these lines. Hence we have a total of n^2 points in front of us. We claim that these are all the points of Π. For let P be any point. If P lies on L, we have already counted it. If P does not lie on L, there is a line through P parallel to L, a line which must be one of $L_2, L_3, \ldots, L_n$. Thus lines $L, L_2, L_3, \ldots, L_n$ together exhaust all the points of Π. Therefore our above count of the points was complete, and there are indeed n^2 points in Π.

Exercises

1. Let Π be the affine plane whose elements are 1, 2, 3, and 4, and whose lines are the six two-element subsets. Prove that any two parallel lines fill up Π.

2. In an affine plane prove that two lines parallel to the same line are either parallel or identical.

3. (This exercise was mentioned in the course of proving Theorem 5.1.) In the notation of the proof of Theorem 5.1, define a function G from M to L, analogous to the function F, and prove that the composite functions FG and GF are both equal to the identity.

4. Let L and M be distinct nonparallel lines in an affine plane of order n. What is the total number of points on $L \cup M$?

5. Let L be a line in an affine plane of order n. What is the total number of points on the lines parallel to L?

6. Prove that any affine plane contains at least four points and at least six lines.

7. Let P be a point in an affine plane of order n. Prove that there are exactly $n + 1$ lines through P.

8. Prove that an affine plane of order n contains exactly $n^2 + n$ lines.

9. Call the set of all lines parallel to a given line a *family* of parallels. Prove that any affine plane of order n contains precisely $n + 1$ families of parallels.

10. Let L be a line in an affine plane and let P be a point not on L. Prove directly (i.e., without using any of the theorems in this section) that the number of lines through P is one larger than the number of points on L. (This can be used as the starting point for an alternate treatment.)

4. Planes of Low Order

The smallest possible order for an affine plane is 2, and it then has four points. Number them as 1, 2, 3, 4. For every two points, say i and j, there has to be a line containing them, and since the line has just two points, it has to consist exactly of i and j. This is true for every method of picking two points from 1, 2, 3, 4:

12, 13, 14, 23, 24, 34

So these are the 6 lines of Π. It is fairly simple to verify that this satisfies all the axioms for an affine plane (compare Exercise 3 of Section 2). In Figure 5.9 the plane is illustrated by a diagram. Note that 13 is parallel to 24, and 12 is parallel to 34, so that the four points constitute a parallelogram. Despite the temptation offered by Figure 5.9, the two diagonals 14 and 23 do not intersect; they are also parallel! The four points constitute a kind of "super-parallelogram." (A sort of model can be pictured by bending one of the diagonals in the third dimension so as to avoid intersecting the other diagonal.)

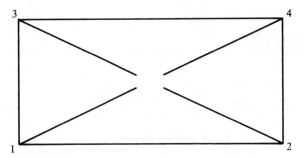

Figure 5.9

The next possible order is 3. By a straightforward discussion we shall show that here, too, there is exactly one possible plane.

Let us take advantage of Theorem 5.2 to arrange the nine points into three parallel lines, as in Figure 5.10. We label the points on the two bottom lines as shown, but delay for a moment labeling the top line.

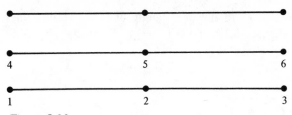

Figure 5.10

Consider next line L joining 1 and 4. It has to have exactly one more point on it and this point cannot be 2, 3, 5, or 6. (Why?) Let the additional point on L be numbered 7. In just the same way, the additional point on the line joining 2 and 5 is a new point, and we number it 8. The line joining 3 and 6 gets the remaining point, number 9. We have reached the setup illustrated in Figure 5.11. Note that the lines 147, 258, and 369 form a second family of three parallel lines.

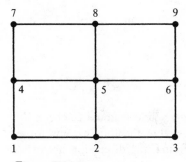

Figure 5.11

From this point on, every step is forced on us. Line 15 has to finish with 9, 26 with 7, and 34 with 8; this is a third family of three parallel lines. It is difficult to furnish an illustration, but Figure 5.12 is an attempt.

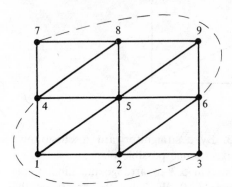

Figure 5.12

The lines 168, 249, 357 are the final family of three parallel lines shown in Figure 5.13. Figures 5.12 and 5.13 ought to be combined but the result probably would not be very helpful.

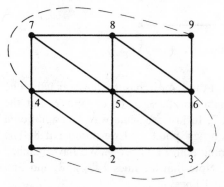

Figure 5.13

We have concluded the construction of this plane of order 3 and we summarize by listing the 12 lines:

123	147	159	168
456	258	267	249
789	369	348	357

It is quite straightforward (although perhaps a little tedious) to check that we do indeed have an affine plane, that is, that the four axioms for an affine plane are satisfied. Up to a renumbering, the discussion above has shown that this is the only affine plane of order 3.

In the next section we shall give a natural way of constructing the plane of order 3 and a similar one of order p for every prime.

Let us see how a discussion in this style evolves for planes of order 4. Harmless renumbering of points brings us to the families of "horizontal" and "vertical" lines of Figure 5.14. Next we try to complete the line joining 1 and 6. There are two possible completions: by points 11 and 16, or by points 12 and 15. It is still a harmless renumbering (amounting to an interchange of the third and fourth vertical lines) to fix the completion as 11, 16. It is now a fact that the remaining lines are completely determined; we shall leave it to an interested reader to see if he can finish the job. We shall say a little more about this plane of order 4 in the next section.

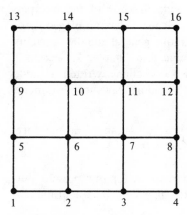

Figure 5.14

Exercises

1. Let A, B, C be three noncollinear points in an affine plane. Prove that there is exactly one point D making $ABCD$ a parallelogram, that is, making BD parallel to AC and CD parallel to AB.

2. (a) In the notation used above for the plane of order three, what is the fourth point needed to complete a parallelogram with the points 1, 3, and 7?
   (b) Is the resulting parallelogram a super-parallelogram; that is, are the diagonals parallel? If not, find the point of intersection of the diagonals.
   (c) Answer (a) and (b) for 1, 4, and 5; also for 1, 5, and 6.

3. A *triangle* in an affine plane is a set of three noncollinear points. How many triangles are there in the affine plane of order 2? In the affine plane of order 3?

4. In the affine plane of order 3, what is the largest number of points with the property that no three are collinear?

5. In the affine plane of order 3, the following theorem is alleged to hold: If *ABC* is a triangle, *D* the remaining point on *AB*, and *E* the remaining point on *AC*, then *BE* and *CD* are parallel. Check this out for two triangles of your choice.

5. Coordinate Affine Planes

In this section we shall construct and discuss affine planes by using coordinates as in elementary analytic geometry.

In the reader's previous study of analytic geometry it was presumably understood (either explicitly or implicitly) that the coordinates were allowed to be any real numbers. We get a better understanding of the interplay between algebra and geometry, and a much richer collection of geometric examples, by taking the coordinates from more general number systems. The exact concept is that of a *field,* as introduced in Chapter 2, Section 3.

We recall that a field is a set with two operations, written as addition and multiplication, which satisfy the standard laws of arithmetic. Important examples are:

1. the field of rational numbers, that is, of numbers of the form a/b where a and b are integers and $b \neq 0$
2. the field of real numbers
3. (This is the example that most interests us for our present purposes). The field Z_p of integers mod p, where p is a prime.

We shall normally use ordinary integers such as $0, 1, \ldots, p - 1$ to represent the elements, with the understanding that addition and multiplication take place modulo p. (Note: Previously we had used the notation $[0], [1], \ldots, [p - 1]$ for the elements of Z_p. In this chapter we shall drop the square bracket because in view of the heavy use of Z_p here, continuation of the bracket would be cumbersome.)

Let F be any field; it might be the real numbers, the rational numbers, the integers mod p, or one of the many other fields that exist. We shall define an *affine plane based on F,* our notation for it being $\Pi(F)$. The points of $\Pi(F)$ are pairs (x, y) of elements of F. The lines of $\Pi(F)$ are described as follows: Given three elements a, b, c of F with a and b not both 0, the set of all (x, y) satisfying the equation $ax + by + c = 0$ is a line.

As various concepts are introduced in this section, we shall illustrate them using the field Z_5 of integers mod 5. We write 0, 1, 2, 3, 4 for the five elements of this field. Addition and multiplication are performed modulo 5. Examples:

$$1 + 3 = 4, \quad 2 + 3 = 0, \quad 4 + 4 = 3, \quad 2 \cdot 2 = 4, \quad 3 \cdot 4 = 2, \quad 4 \cdot 4 = 1$$

There are 25 points in $\Pi(Z_5)$. We exhibit them in a square array (see Figure 5.15):

(0, 4)	(1, 4)	(2, 4)	(3, 4)	(4, 4)
(0, 3)	(1, 3)	(2, 3)	(3, 3)	(4, 3)
(0, 2)	(1, 2)	(2, 2)	(3, 2)	(4, 2)
(0, 1)	(1, 1)	(2, 1)	(3, 1)	(4, 1)
(0, 0)	(1, 0)	(2, 0)	(3, 0)	(4, 0)

Figure 5.15

A typical example of an equation of a line is $2x + 3y + 1 = 0$. We can find the points on it systematically by setting $x = 0, 1, \ldots, 4$ in turn and solving for y. To do the arithmetic efficiently we can solve algebraically for y. Multiply $2x + 3y + 1 = 0$ by 2 to get $4x + y + 2 = 0$, or $y = -4x - 2$, or finally $y = x + 3$. (Remember that we are working modulo 5.) The points are then obtainable rapidly from the following table:

x	$y = x + 3$
0	3
1	4
2	0
3	1
4	2

If we select these five points from Figure 5.15, we get a line that can be pictured as broken when it reaches the top of the tableau; in Figure 5.16 we have inserted a dashed line to join the two "parts" of the broken line.

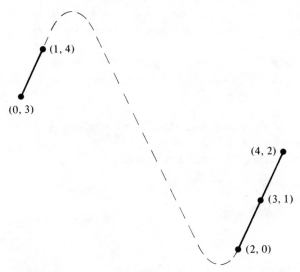

Figure 5.16

We return to $\Pi(F)$ with F a general field. The verification that $\Pi(F)$ is an affine plane is fairly simple algebra, but we have some choice in the techniques to be used. We shall arrange the work in the "slope-intercept" form. Although this calls for some case distinctions, it has the advantage of pinning the lines down in a concrete way, and probably fitting well with the reader's previous study of analytic geometry.

One might say that the main source of difficulty is that the equation of a line in the form $ax + by + c = 0$ is not unique; if we multiply the equation by a nonzero constant, we get the same line. To make the equation unique we choose to solve it for y, but we have to separate the case $b = 0$, where it is impossible to solve for y. When $b = 0$, the equation is $ax + c = 0$ with $a \neq 0$, or $x = -c/a$. We are thus led to speak of vertical and nonvertical lines.

A *vertical line* is given by an equation of the form $x = d$, with d in F. Illustration: With $F = Z_5$ and $d = 3$ we get the vertical line $x = 3$ consisting of the five points $(3, 0)$, $(3, 1)$, $(3, 2)$, $(3, 3)$, and $(3, 4)$.

A *nonvertical line* is given by an equation of the form $y = mx + e$ with m and e in F. We call m the *slope* and e the *y-intercept*. An illustration in Z_5 was given above; we converted the line $2x + 3y + 1 = 0$ to the slope-intercept form $y = x + 3$. Thus the slope is 1 and the y-intercept is 3 in this example.

(Note: The letter b is customary for the y-intercept; we switched to e to avoid confusion with the use of b in $ax + by + c = 0$.)

We now investigate systematically the intersection of two lines, and the discussion divides into three cases, with the third case splitting further into three subcases.

CASE 1.   *Two vertical lines.* Let the lines be $x = d$ and $x = d'$. Of course, if $d = d'$, the lines are identical. If $d \neq d'$, the lines do not intersect.

CASE 2.   *A vertical line and a nonvertical one.* Let the lines be $x = d$ and $y = mx + e$. Obviously there is a unique solution, namely $x = d$, $y = md + e$.

CASE 3.   *Two nonvertical lines.* We take them to be $y = mx + e$ and $y = m'x + e'$. We make further distinctions.

3(a). $m \neq m'$. Subtracting the two equations we get

$$(m - m')x + e - e' = 0$$

Since $m - m' \neq 0$ we can solve for x:

(1) $x = -\dfrac{e - e'}{m - m'}$

We substitute for x in $y = mx + e$ to get the value of y:

(2) $y = -m\dfrac{e - e'}{m - m'} + e$

$ = \dfrac{me' - em'}{m - m'}$

Formulas (1) and (2) give the unique point of intersection of the two lines.

3(b). $m = m'$, $e = e'$. The two lines are identical.

3(c). $m = m'$, $e \neq e'$. Subtracting the two equations yields the impossible statement $e - e' = 0$, so the two lines do not intersect.

We summarize what we have proved in a theorem:

Theorem 5.4.   Let F be any field. In $\Pi(F)$ define points, lines (vertical and nonvertical), and slope as above. Then: Two distinct vertical lines do not intersect. A vertical line and a nonvertical line intersect in one point. Two nonvertical lines with different slopes intersect in one point. Two distinct nonvertical lines with the same slope do not intersect.

In all cases of Theorem 5.4 we see that two distinct lines intersect either in one point or in no points at all. This accomplishes part of the task of proving Axiom 1 for affine planes: If two distinct points can be joined by a line, that line is unique.

To complete the job of verifying Axiom 1 we need to prove that there is a line joining any two points. We do that in Theorem 5.6, after proving Theorem 5.5. The setup in Theorems 5.5–5.7 is as in Theorem 5.4.

Theorem 5.5.   Let (x_1, y_1) be any point in $\Pi(F)$, m any element of F. There exists a unique line through (x_1, y_1) with slope m, and an equation for it is

(3) $y - y_1 = m(x - x_1)$

Proof.   Equation (3) is of the right form to determine a line. It is satisfied by (x_1, y_1). When we put it in the form

$y = mx + (y_1 - mx_1)$

we recognize that the slope is m (and the y-intercept is $y_1 - mx_1$). The uniqueness of the line follows from the last sentence of Theorem 5.4.

Theorem 5.6.   There exists a line joining any two distinct points of $\Pi(F)$.

Proof.   Let the points be (p, q) and (r, s). We distinguish two cases. If $p = r$, the vertical line $x = p$ will do. If $p \neq r$, we define

(4) $m = \dfrac{q - s}{p - r}$

and take the line through (p, q) with slope m, which, by Theorem 5.5 is

(5) $y - q = m(x - p)$

By direct substitution we see that this line goes through (r, s).

We remark again that in Theorem 5.4 we have already proved the uniqueness of the line obtained in Theorem 5.6.

We add a further note concerning Theorem 5.6, justifying retrospectively formula (4).

Theorem 5.7.   Let (p, q) and (r, s) be points of $\Pi(F)$ with $r \neq p$. Then the line joining them is nonvertical and its slope is given by $(q - s)/(p - r)$.

Proof.   In proving Theorem 5.6 we identified the line in question as (5), and this equation shows that the slope is m, with m given by (4).

We are ready for the main theorem.

Theorem 5.8.   Let F be any field, and let $\Pi(F)$ be the set of all pairs of elements of F, with lines defined as above. Then $\Pi(F)$ is an affine plane.

Proof.   Axioms 3 and 4 are quite obvious. With Theorems 5.4 and 5.6 we have verified Axiom 1. It remains to check Axiom 2.

Suppose L is a given line and (p, q) a point not on L. We have to show that there is a unique line through (p, q) parallel to L. If L is vertical, then (Theorem 5.4) the line we want is necessarily the vertical line $x = p$. If L is nonvertical, say with slope m, then (again Theorem 5.4) the line we want goes through (p, q) with slope m. By Theorem 5.5, there is exactly one such line.

For every prime p the field Z_p gives us an affine plane $\Pi(Z_p)$ of order p. Of course for $p = 2$ and 3 we get the planes of orders 2 and 3 discussed in the preceding section.

Any finite fields other than the fields Z_p will give us more examples of affine planes. It is beyond the scope of this book to discuss finite fields, but we shall record the facts. The number of elements in a finite field has to be a power of a prime. Any power of a prime will do, and there is exactly one field having that many elements.

The first such field, other than the fields Z_p, has four elements. Note that this field of order 4 is *not* the ring of integers mod 4. (Why?)

Without explaining where it all comes from, we give addition and multiplication tables for the field of four elements. We take the elements to be 0, 1, u, v.

Addition

	0	1	u	v
0	0	1	u	v
1	1	0	v	u
u	u	v	0	1
v	v	u	1	0

Multiplication

	0	1	u	v
0	0	0	0	0
1	0	1	u	v
u	0	u	v	1
v	0	v	1	u

We conclude this section by presenting in tabular form the known facts concerning affine planes of order n up to $n = 11$. We shall return to this in Section 7.

ORDER

2   The only plane is: $\Pi(Z_2)$. (See Section 4.)

3   The only plane is: $\Pi(Z_3)$. (See Section 4.)

4   The only plane is the one based on the field of four elements. This is a little harder to prove than the argument for order 3.

5   The only plane is $\Pi(Z_5)$. This is a little harder still.

6   There are no planes of order 6, as we shall prove in Section 7.

7   The only plane is $\Pi(Z_7)$. By optimal organization of the arguments, case distinctions can be minimized and the proof given in about half a dozen pages.

8   Only the plane based on the field of eight elements exists. This has been verified only with the aid of a large computer.

9   In addition to $\Pi(F)$ where F is the field of nine elements, three more planes are known. It is not known whether more exist.

10   It is unknown whether any exist.

11   It is unknown whether any exist in addition to $\Pi(Z_{11})$.

Exercises

In Exercises 1–3 the points and lines lie in $\Pi(Z_5)$.

1. Find the point of intersection of lines $2x + y = 3$, $-x + 3y = 2$.

2. Find the equation of the line through $(2, 1)$ parallel to $x - 2y = 1$.

3. For what values of k are lines $(k + 1)x - y = 2$, $(k + 2)x + ky = 3$ parallel?

4. With F the field of four elements described above, number the 16 points of $\Pi(F)$ as follows:

13	14	15	16
$(0, v)$	$(1, v)$	(u, v)	(v, v)
9	10	11	12
$(0, u)$	$(1, u)$	(u, v)	(v, u)
5	6	7	8
$(0, 1)$	$(1, 1)$	$(u, 1)$	$(v, 1)$
1	2	3	4
$(0, 0)$	$(1, 0)$	$(u, 0)$	$(v, 0)$

List 5 of the 20 lines explicitly. Compare with the discussion at the end of Section 4.

6. Parallelograms and Midpoints

The modest scope of this chapter precludes any attempt to develop a substantial segment of geometry. We shall confine ourselves in this section to a brief investigation of parallelograms and the way their diagonals intersect.

We have already mentioned parallelograms in Section 4, in connection with the plane of order 2, and in Exercises 1 and 2 of that section. But let us repeat the definition. Given four distinct points A, B, C, D (note that the order matters) we say that $ABCD$ is a *parallelogram* if AB is parallel to CD and AC is parallel to BD (Figure 5.17). We call lines AD and BC the diagonals of the parallelogram.

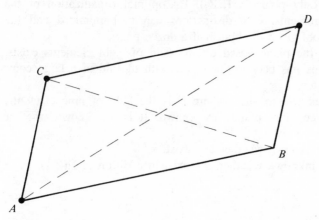

Figure 5.17

We ask whether the diagonals intersect, and if so, what is the nature of the point of intersection? To gain some insight into the question, we try it for a coordinate affine plane. In doing this we shall assume point A to be the origin $(0, 0)$. For B and C we assume general coordinates (p, q) and (r, s). We seek what D has to be in order to make $ABCD$ a parallelogram. The distinction we have made between vertical and nonvertical lines gives us some extra trouble here. We shall carry out the argument on the assumption that AB and AC are nonvertical, leaving it as an exercise (Exercise 1) to see that we get the same answer if either is vertical.

Write (x, y) for D. The slope of AB is q/p; the slope of CD is $(y - s)/(x - r)$. The requirement that AB and CD be parallel means (Theorem 5.4) that these two slopes are equal.

$$(6) \quad \frac{q}{p} = \frac{y - s}{x - r}$$

In the same way the parallelism of AC and BD gives us

$$(7) \quad \frac{s}{r} = \frac{y - q}{x - p}$$

We ask the reader to carry out the algebra of solving (6) and (7). The answer is $x = p + r$, $y = q + s$, and we record these coordinates in Figure 5.18. (A reader who has some familiarity with vectors will recognize in Figure 5.18 the parallelogram law for adding vectors. There are numerous examples in physics of vectors—velocities, accelerations, forces —which add in this way.)

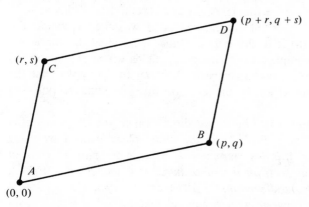

Figure 5.18

When are diagonals AD and BC parallel? We investigate this, again leaving it to the reader to check out the exceptional possibility that one or both of these lines is vertical. The slope of AD is $(q + s)/(p + r)$; that of BC is $(q - s)/(p - r)$. The condition for equality is

$$\frac{q+s}{p+r} = \frac{q-s}{p-r}$$

or

$$(p-r)(q+s) = (p+r)(q-s)$$

or

$$pq + ps - rs - rq = pq - ps + rq - rs$$

or finally

$$2(ps - rq) = 0$$

Now $ps - rq = 0$ means that points A, B, C are collinear. (Why?) We are, of course, ruling this out for we want $ABCD$ to be a genuine parallelogram. So we appear to have proved that AD and BC are not parallel. But wait—there is a bizarre possibility that we must not ignore, for we are allowing F to be any field. It is conceivable that $2 = 0$, as is certainly the case for the field Z_2 of integers mod 2.

Let us pause to explain this a little more. In any field F we have a unit element that we write as 1. We may add 1 to itself as many times as we like. Now there are two possibilities. In the familiar or ordinary case (i.e., fields such as rational numbers or real numbers) the result

$$(8) \quad 1 + 1 + \cdots + 1 \qquad (n \text{ terms})$$

of adding 1 to itself n times never gives 0, no matter what the positive integer n is. On the other hand, it may happen that expression (8) does get to be 0, as is the case for the fields Z_p, and certain other fields as well.

It happens (the argument is simple) that if we take the *smallest* n such that expression (8) is 0, this n is a prime; it is called the *characteristic* of the field. To complete the terminology, fields where expression (8) is never 0 are said to have characteristic 0. Note, for example, that the field of four elements given at the end of the preceding section has characteristic 2.

We can now summarize the above discussion of the diagonals of a parallelogram. Let F be a field, $\Pi(F)$ the affine plane defined by using coordinates in F. Then: *if F has characteristic $\neq 2$, in any parallelogram of $\Pi(F)$ the diagonals intersect; on the other hand, if F has characteristic 2, the diagonals of any parallelogram of $\Pi(F)$ are parallel.* (In Section 4 we called such a parallelogram a "super-parallelogram.")

We emphasize at this point that we have been discussing coordinate affine panes. What happens if we examine the parallelograms of a general affine plane? Answer: The worst is possible. It is conceivable that some parallelograms are super and some are not, in a single fixed affine plane! Clearly, such a plane Π could not be of the coordinate form $\Pi(F)$ for any field F; it is customary then to say that the introduction in Π of coordinates from a field is impossible. It is beyond the scope of our treatment to exhibit such an example.

There is however a beautiful theorem, due to Gleason, which says, essentially, that if every parallelogram in a finite affine plane Π is super, then coordinates can be introduced in Π; that is, Π can be identified with $\Pi(F)$ for a suitable field F (necessarily F would have characteristic 2). It is not known whether or not Gleason's theorem is true in the infinite case.

We return to the parallelogram of Figure 5.18 in the coordinate affine plane $\Pi(F)$. Assume that F has characteristic $\neq 2$, so that diagonals AD and BC intersect. In what point do they intersect? We leave it as an exercise (2) to check that the answer is

$$(9) \quad \left(\frac{p+r}{2}, \frac{q+s}{2}\right)$$

Note that since the characteristic of F is not 2, it is legal to put 2 in the denominator.

The form of (9) leads us to discuss midpoints. In a coordinate affine plane $\Pi(F)$, with the characteristic of F different from 2, we can simply decree that point (9) is the midpoint of the points (p, q) and (r, s). It is easy enough to check that point (9) lies on the line joining (p, q) and (r, s). But we do not attempt to say that (9) is equidistant from the two points, for in our geometry there is no notion of distance, and we repeat that the concept of midpoint is being treated by outright definition.

This is all very well in a coordinate affine plane. Can we talk about midpoints in any affine plane? We certainly do not have a formula such as (9) to fall back on. However, we can use the parallelogram idea to approach the matter in a devious way. Let points A and D be given in an affine plane. As a device for getting a midpoint P of AD, drape a parallelogram $ABCD$ around AD (Figure 5.19) and take P to be the intersection of diagonals AD and BC (it being assumed that we did not run into

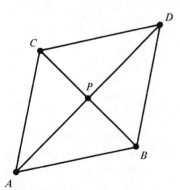

Figure 5.19

a super-parallelogram). But now the crucial question is this: Suppose a different parallelogram were used, would we get the same *P*?

We hasten to inform the reader that the answer can be "no," and we add that any example of this must also be an example of an affine plane that cannot admit coordinates from a field.

We shall present a famous example, due to Moulton.

> F. R. Moulton (1872–1952) was on the faculty of the University of Chicago (in both astronomy and mathematics) from 1898 to 1926. In collaboration with T. C. Chamberlin, he formulated the famous planetesimal hypothesis for the origin of the solar system. His popular account of astronomy (*Consider the Heavens,* 1935) makes stimulating reading to this day. However, there is a passage on page 107 which might serve as a warning not to be too daring in one's predictions.
>
> "Yet in fairness to those who by training are not prepared to evaluate the fundamental difficulties of going from one planet to another, or even from the earth to the moon, it must be stated that there is not the slightest possibility of such journeys."

In Moulton's example the number of points is infinite, and in fact the points are the same as the points of the ordinary Euclidean plane (pairs of real numbers). Many of the lines are also ordinary Euclidean lines, and the others are just a little distorted.

Here are the precise details. The lines of the Euclidean plane that are horizontal, or vertical, or have negative slope, are carried over to the Moulton plane unchanged. Lines of positive slope are broken as they cross the *x*-axis. Above the *x*-axis their slopes change to a smaller value, which is smaller by a fixed factor. To make the example explicit and as simple as possible, we take the fixed factor to be $\frac{1}{2}$. Figure 5.20 illustrates three broken lines of the Moulton plane.

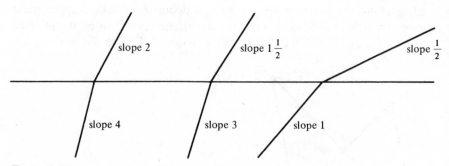

Figure 5.20

It is elementary and straightforward, but a little lengthy, to verify that the Moulton plane is indeed an affine plane. We invite an ambitious reader to take a shot at the details. As a little guidance we mention two crucial points. Figure 5.21 shows what two parallel broken lines look like, the point being that the two halves themselves must be parallel as ordinary lines. In fact, if we make an agreement to use only the slope below the *x*-axis, we can simply say that two lines are parallel if and only if they have the same slope.

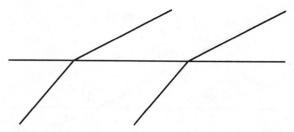

Figure 5.21

If we have two broken lines of unequal slope, they meet in exactly one point. Figure 5.22 illustrates the two cases; in the left half of the figure the steeper line meets the *x*-axis to the left of the less steep one, while in the right half of the figure it is the opposite. In the first case the lines intersect below the *x*-axis, and in the second case they intersect above the *x*-axis.

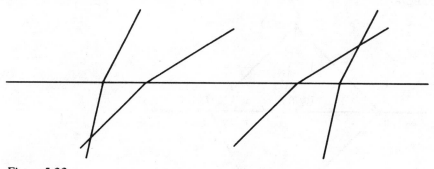

Figure 5.22

Let us check out whether, in the Moulton plane, the parallelogram method for getting a midpoint yields a unique result. For our points A and D we choose $(-1, 0)$ and $(1, 0)$, and our first parallelogram (Figure 5.23) is such that all the lines involved are ordinary Euclidean lines (horizontal, vertical, or with negative slope). So with this version the proposed midpoint turns out to be $(0, 0)$ as usual. In Figure 5.24 we exhibit another choice of a parallelogram draped around points A and D. Despite its odd-looking shape it *is* a parallelogram. The slope of AC is 1, which is half the slope, namely 2, of BD, and so AC and BD are parallel. When the details are worked out, we find that diagonals AD and BC intersect in the "midpoint" ($\frac{1}{7}$, 0). So, the Moulton plane cannot admit coordinates from a field.

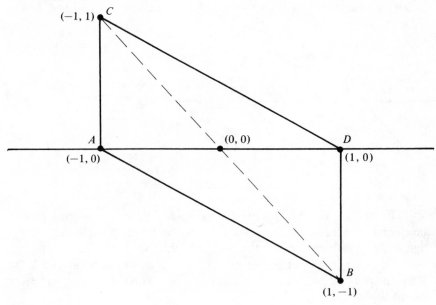

Figure 5.23

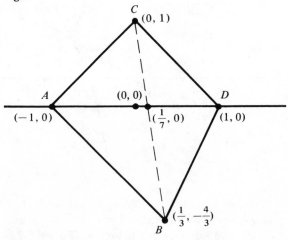

Figure 5.24

Because we have placed so much emphasis on the uniqueness of the midpoint construction by parallelograms, it is natural to ask: If a plane Π does satisfy this uniqueness requirement, can it be identified with $\Pi(F)$ for a suitable field F? This is an excellent question; however a full answer would take us too far afield. But we can sketch the facts. (There is one slight inaccuracy in the following statements; we ought to be dealing with objects called "projective planes," which are slightly different from affine planes.)

1. For Π to admit coordinatization from a field we have to assume Pappus' theorem. This theorem is illustrated in Figure 5.25: If *A*, *B*, *C* are collinear, *D*, *E*, *F* are collinear, and *X*, *Y*, *Z* are the three "cross-joins," then *X*, *Y*, *Z* are collinear.

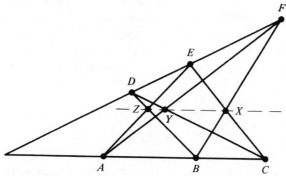

Figure 5.25

2. The theorem of Desargues is illustrated in Figure 5.26; let triangles *ABC* and *DEF* be in perspective from 0 (i.e., *AD*, *BE*,

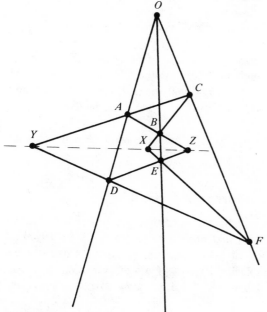

Figure 5.26

and *CF* meet at 0), and suppose that *BC* and *EF* meet at *X*, *CA* and *FD* meet at *Y*, and *AB* and *DE* meet at *Z*. Then *X*, *Y*, and *Z* are collinear. The assumption that Π satisfies Desargues' theorem turns out to be equivalent to the possibility of introducing coordinates from a field, except that we discard the commutative law of multiplication.

3. The assumption that the midpoint obtained from parallelograms is independent of the choice of parallelogram is equivalent again to coordinatization, where this time not only is commutativity of multiplication discarded, but associativity is replaced by a weaker axiom called the "alternative" law. The alternative law (really a pair of axioms) asserts that the associative law holds when there is a repeated letter:

$$a(bb) = (ab)b, \quad (bb)a = b(ba)$$

Exercises*

1. In a coordinate affine plane let $A = (0, 0)$, $B = (p, q)$, $C = (r, s)$ and suppose they are not collinear. Prove that, in the remaining cases where *AB* or *AC* is vertical, point *D* making *ABCD* a parallelogram is again given by $D = (p + r, q + s)$.

2. Let the notation be as in Exercise 1, and assume the underlying field does not have characteristic 2. Prove that *AD* and *BC* intersect at $((p + r)/2, (q + s)/2)$.

3. Prove that in the Moulton plane no parallelogram is a superparallelogram.

4. Let Π be the affine plane over a field of characteristic $\neq 2$ or 3. Let *ABC* be any triangle in Π. If *D* is the midpoint of *BC*, *E* the midpoint of *CA*, and *F* the midpoint of *AB*, lines *AD*, *BE*, and *CF* are called the *medians*. Prove that the three medians meet in a point.

5. In Exercise 4, what can you say about the medians if the characteristic is 3?

7. The Nonexistence of Planes of Order 6

In this final section we shall give a complete proof that no affine plane can have order 6. This was first proved by G. Tarry (1843–1913) in 1901 by a brutal enumeration of possible cases. Actually, Tarry proved something stronger. The technical statement of his theorem is as follows: There cannot exist two orthogonal 6 by 6 Latin squares; roughly speaking,

---

*Exercises 1 and 2 complete the discussion of parallelograms in the text.

this means that in a certain sense we cannot even start the construction of an affine plane of order 6.

At the end of this section we shall indicate that the proof really works to show the nonexistence of affine planes of order n for a whole collection of integers n starting at 6. In this more general form the theorem is due to Bruck and Ryser. Subsequently the proof was simplified by Chowla and Ryser. Our account will follow closely the account of the Chowla-Ryser proof given by Marshall Hall in his book *The Theory of Groups* (New York: Macmillan, 1959, pp. 394–397).

We imagine the proof will strike most readers as very remarkable and also strange. What was the origin of the idea of using all these variables, and combining them in these intricate linear and quadratic expressions? This is not an easy question to answer. One way of describing what happened is that there is here a fusion of algebra and geometry, reminiscent of the way Descartes united algebra and geometry when he invented analytic geometry. The algebraic techniques arose in other contexts, and were available at the moment of inspiration when Bruck and Ryser saw their applicability.

Theorem 5.9.   There is no affine plane of order 6.

Proof.   Suppose we have on hand an affine plane Π of order 6, and we eventually reach a contradiction.

The first step is to invent 36 variables, one for each of the 36 points in Π. We call them $x_1, x_2, \ldots, x_{36}$. It is to be noted that we are not yet paying any attention to the way these 36 points are grouped into lines; in other words, the order in which we list the points as $x_1, x_2, \ldots, x_{36}$ is quite random. It will be handy (although we must admit a little careless) to use the same letter x_i for a point of Π and for the variable representing it.

We know that Π has 42 lines (Exercise 8 in Section 3). We write them as $L_1, L_2, \ldots, L_{42}$, again listing them haphazardly in any order. Each of these lines has 6 points on it. We have no knowledge of what points go on what line. Just to illustrate: It might be that the points on L_1 and L_2 are given in the following lists.

(10) $L_1: x_2, x_5, x_6, x_{27}, x_{33}, x_{35}$
        $L_2: x_{19}, x_{20}, x_{28}, x_{31}, x_{34}, x_{36}$

Now we come to a key idea: To indicate in a convenient way the connection between points and lines we decide to add up the 6 variables that occur in L_1:

(11) $x_2 + x_5 + x_6 + x_{27} + x_{33} + x_{35}$

Again making a careless but convenient identification, we write L_1

both for the line in question and (with the small change about to be made) for expression (11).

The small change just mentioned would probably best be motivated by experimenting with expression (11) and the ones for the other lines, and seeing that something more is needed. However we merely say this: Parallel lines and their properties are so important in the axioms for an affine plane that we ought to pay more attention to them. We recall (Exercise 9 in Section 3) that the 42 lines of Π fall into 7 families, each containing 6 parallel lines. We might think of each family of 6 parallel lines as a direction. We wish to record, in addition to the 6 points a given line has, what its direction is. So we decide to add 7 more variables, one for each direction, and we name them x_{37}, x_{38}, . . . , x_{43}. Figure 5.27 may be useful in picturing the meaning of these new variables.

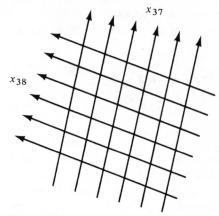

Figure 5.27

We continue the illustration in (10) of L_1 and L_2. Suppose for instance that the direction of L_1 is x_{42}, and that the direction of L_2 is x_{39}. Then we write

(12) $$L_1 = x_2 + x_5 + x_6 + x_{27} + x_{33} + x_{35} + x_{42}$$
$$L_2 = x_{19} + x_{20} + x_{28} + x_{31} + x_{34} + x_{36} + x_{39}$$

Before the algebraic maneuvers start, we have one final adjustment which we motivate (perhaps a little lamely) by the observation that we now have 43 x's (36 points and 7 directions), and only 42 lines. To redress the balance, we invent a 43rd line (a ghost line, so to speak) and set it equal to the sum of the 7 directions:

(13) $L_{43} = x_{37} + x_{38} + x_{39} + x_{40} + x_{41} + x_{42} + x_{43}$

Let H be the sum of the squares of $L_1, L_2, \ldots, L_{42}, L_{43}$; that is

(14) $H = L_1{}^2 + L_2{}^2 + \cdots + L_{42}{}^2 + L_{43}{}^2$

We expand the terms $L_1{}^2$, $L_2{}^2$, $L_{43}{}^2$ occurring in (14) in order to make the expression for H more concrete. By (12) and (13) we have

$$H = (x_2 + x_5 + x_6 + x_{27} + x_{33} + x_{35} + x_{42})^2 + (x_{19} + x_{20} + x_{28}$$
$$+ x_{31} + x_{34} + x_{36} + x_{39})^2 + \cdots + (x_{37} + x_{38} + x_{39} + x_{40} + x_{41}$$
$$+ x_{42} + x_{43})^2$$

It turns out that the axioms for an affine plane enable us actually to evaluate H, and we find a delightfully simple expression for it.

When term L_i gets squared we find a combination of terms such as $x_1{}^2$ and $x_1 x_2$ (squares of x's and products of two distinct x's). We investigate the squared terms first. How often will $x_1{}^2$ appear in H? The answer is that we get a contribution $x_1{}^2$ every time x_1 occurs on a line. Because in Π there are 7 lines through every point (Exercise 7 in Section 3) we get a total of $7x_1{}^2$. The argument and conclusion are the same for any of $x_2, \ldots, x_{36}$. For a direction such as x_{37}, we have to reconsider. We will get a term $x_{37}{}^2$ from any line having the direction x_{37}, and there are six of these. But in addition we obtain $x_{37}{}^2$ when we square the ghost line L_{43}. Together, these two pieces give us the coefficient 7 for $x_{37}{}^2$, the same as for $x_1{}^2, \ldots, x_{36}{}^2$. This pleasant "coincidence" shows that we uniformly get the coefficient 7 for every squared term.

We turn to the cross-product terms. Note, for instance, that when L_1 is squared there is a cross-product contribution from x_2 and x_5, and it is $2x_2 x_5$. Thus when we think of a product term as occurring just once we mean that it occurs with a coefficient 2. Now we claim that every cross-product term occurs just once. There are three different arguments, the typical cases being $x_1 x_2$, $x_1 x_{37}$, and $x_{37} x_{38}$, that is, two points, a point and a direction, and two directions.

$x_1 x_2$.   This can arise only from a line containing x_1 and x_2, and there is exactly one such line.

$x_1 x_{37}$.   The only source is a line through x_1 with direction x_{37} and (by the parallel postulate) there is exactly one such.

$x_{37} x_{38}$.   This term arises only from the ghost line L_{43}.

To summarize, we have proved the following equation:

(15) $H = L_1{}^2 + L_2{}^2 + \cdots + L_{43}{}^2$
$$= 7(x_1{}^2 + x_2{}^2 + \cdots + x_{43}{}^2) + 2(x_1 x_2 + x_1 x_3 + \cdots + x_{42} x_{43})$$

In the rest of the proof we make three successive changes of variables that combine to reduce (15) to a form in which we recognize an impossibility.

The first change of variable is based on the idea that cross-product terms are awkward to handle, and we therefore "complete the square" to get rid of them. Probably the obvious choice for this purpose would be

(16) $(x_1 + x_2 + \cdots + x_{43})^2$

for (16) would account precisely for all product terms in (15). However, it would still be necessary to account for the square terms. After some experimentation, one finds it advisable to drop x_1 in (16), so as to allow some room for maneuver in fitting in the square terms. So, to start the change of variables, we set

(17) $y_1 = x_2 + x_3 + \cdots + x_{43}$

What terms involving x_2, other than those already present in y_1^2, are needed if we are to get H? The usual process of completing the square leads us to rewrite this as

$$6\left(x_2 + \frac{x_1}{6}\right)^2 - \frac{x_1^2}{6}$$

We consequently set $y_2 = x_2 + x_1/6$ and similarly we set

(18) $y_i = x_i + \dfrac{x_1}{6}$ $(i = 2, \ldots, 43)$

It is pleasant to find

(19) $H = y_1^2 + 6(y_2^2 + \cdots + y_{43}^2)$

To verify (19) we note that we constructed the y's so as to satisfy (19) except possibly for the term x_1^2. But x_1^2 does not occur in y_1^2. In each of the 42 terms $y_2^2, \ldots, y_{43}^2$ we get a contribution $x_1^2/6$, so the net total is $42(x_1^2/6) = 7x_1^2$, as required.

Before we proceed to the second change of variable, we make an observation showing how to switch our problem from the x's to the y's. Note that the y's are linear combinations of the x's with coefficients that are rational numbers; this is quite obvious from (17) and (18). Conversely, we can solve backward for the x's in terms of the y's, again getting linear combinations with rational coefficients. To see this we add the equations in (18) and then subtract (17); the result is

$$7x_1 + y_1 = y_2 + \cdots + y_{43}$$

which gives us x_1 in terms of the y's. Then the equations in (18) give us the remaining x's.

Summarizing up to this point, we have

(20) $H = L_1^2 + \cdots + L_{43}^2 = y_1^2 + 6(y_2^2 + \cdots + y_{43}^2)$

with each L_i a linear combination of the y's with rational coefficients. Our aim is to show that (20) is impossible.

The difficulty we face in trying to work with (20) is the abundance of 6's on the right-hand side. We now change almost all the 6's to 1's by a device (about which we shall say a little more at the end of the discussion). Here is an identity which the reader should carefully check:

$$(21) \quad 6(a^2 + b^2 + c^2 + d^2) = (2a + b + c)^2 + (2b - a + d)^2$$
$$+ (2c - a - d)^2 + (2d - b + c)^2$$

Inspired by (21), we set

$$(22) \quad \begin{aligned} z_2 &= 2y_2 + y_3 + y_4 \\ z_3 &= -y_2 + 2y_3 + y_5 \\ z_4 &= -y_2 + 2y_4 - y_5 \\ z_5 &= -y_3 + y_4 + 2y_5 \end{aligned}$$

Then

$$(23) \quad (z_2^2 + z_3^2 + z_4^2 + z_5^2) = 6(y_2^2 + y_3^2 + y_4^2 + y_5^2)$$

We continue in groups of four with expressions analogous to (23) for z_6, z_7, z_8, z_9 in terms of y_6, y_7, y_8, y_9, and so on until we reach $z_{38}, z_{39}, z_{40}, z_{41}$ expressed in the same way in terms of $y_{38}, y_{39}, y_{40}, y_{41}$. To make the switch of variables complete we harmlessly set $z_1 = y_1$, $z_{42} = y_{42}$, $z_{43} = y_{43}$. Then from (20) and the 10 equations like (23) we get

$$(24) \quad z_1^2 + \cdots + z_{41}^2 + 6(z_{42}^2 + z_{43}^2) = y_1^2 + 6(y_2^2 + \cdots + y_{43}^2)$$

It is to be noted that the right-hand side of (24) is the expression H that we are studying.

We need a remark analogous to the one made after the first change of variable: that we can solve for the y's backwards in terms of the z's, and that the result serves to express the y's as linear combinations of the z's with rational coefficients. While the exact expressions do not matter, we might as well exhibit explicitly the result of solving (22) for the y's.

$$\begin{aligned} 6y_2 &= 2z_2 - z_3 - z_4 \\ 6y_3 &= z_2 + 2z_3 - z_5 \\ 6y_4 &= z_2 + 2z_4 + z_5 \\ 6y_5 &= z_3 - z_4 + 2z_5 \end{aligned}$$

We use (24) to switch (20) to

$$(25) \quad H = L_1^2 + \cdots + L_{43}^2 = z_1^2 + \cdots + z_{41}^2 + 6(z_{42}^2 + z_{43}^2)$$

and observe that each L_i is now a linear combination of the z's with rational coefficients. Our objective has now become to prove (25) untenable.

The final switch we make is an elimination of variables rather than a change of variables. Suppose that in (25) we set z_1 equal to a linear combination with rational coefficients of $z_2, \ldots, z_{43}$. The result will be an equation asserting equality between two expressions in $z_2, \ldots, z_{43}$. If we find this new equation to be impossible, we have the contradiction we are seeking. Now, we are going to choose the expression for z_1 in terms of $z_2, \ldots, z_{43}$ in such a way that L_1^2 and z_1^2 cancel. The result will be simply to cancel L_1^2 and z_1^2 in (25), at the expense of switching $L_2, \ldots, L_{43}$ to new linear combinations just involving the letters $z_2, \ldots, z_{43}$. We shall repeat this device 41 times!

We turn to the details. We began the discussion with an illustration where L_1 was given by (12). After two changes of variable we have almost lost track of the form of L_1 (and anyway (12) was only an example). All we know for sure is that L_1 has the form

$$L_1 = az_1 + bz_2 + \cdots + cz_{43}$$

where the coefficients $a, b, \ldots, c$ are rational numbers. We set

(26) $az_1 + bz_2 + \cdots + cz_{43} = \pm z_1$

If a is neither 1 nor -1, we can use either sign in (26). Let us say we take the plus sign. Then we can solve for z_1, getting

$$(1 - a)z_1 = bz_2 + \cdots + cz_{43}$$

or

(27) $z_1 = \dfrac{bz_2 + \cdots + cz_{43}}{1 - a}$

If $a = -1$, this will do fine, but if $a = 1$, we switch to $-z_1$ in (26) and then in place of (27) we get

(28) $z_1 = -\frac{1}{2}(bz_2 + \cdots + cz_{43})$

We return to (25) and in it we replace z_1 by the right side of (27) or (28), every time that z_1 occurs in (25). By the way in which we set up (27) and (28), L_1^2 and z_1^2 will cancel. No further z_1's occur on the right side of (25). On the left side of (25) we may have z_1 appearing in some or all of $L_2, \ldots, L_{43}$. Every time it does, we replace it. The result will be to replace $L_2, \ldots, L_{43}$ by rational linear combinations of $z_2, \ldots, z_{43}$. It is convenient to continue writing $L_2, \ldots, L_{43}$ for these new expressions. With this understood, (25) has now become

(29) $L_2^2 + L_3^2 + \cdots + L_{43}^2 = z_2^2 + z_3^2 + \cdots + z_{41}^2 + 6z_{42}^2 + 6z_{43}^2$

We repeat the process, setting $L_2 = \pm z_2$ (using the choice of sign to avoid a clash if the coefficient of z_2 in L_2 is ± 1), solving for z_2 in terms of $z_3, \ldots, z_{43}$ and substituting in (29). We reach

$$L_3{}^2 + \cdots + L_{43}{}^2 = z_3{}^2 + \cdots + z_{41}{}^2 + 6z_{42}{}^2 + 6z_{43}{}^2$$

where $L_3, \ldots, L_{43}$ are now rational linear combinations of $z_3, \ldots, z_{43}$.

In all we patiently repeat the process 41 times, after which the crucial equation has shrunk to

$$L_{42}{}^2 + L_{43}{}^2 = 6z_{42}{}^2 + 6z_{43}{}^2$$

Here L_{42} and L_{43} are rational linear combinations of z_{42} and z_{43}. To look at this explicitly, suppose

$$L_{42} = pz_{42} + qz_{43}, \; L_{43} = rz_{42} + sz_{43}, \; p, \, q, \, r, \, s$$

being rational numbers. Thus

$$(30) \quad (pz_{42} + qz_{43})^2 + (rz_{42} + sz_{43})^2 = 6z_{42}{}^2 + 6z_{43}{}^2$$

Now (30) is an identity; that is, if the left side is rearranged by squaring and gathering terms, we must get exactly the same thing as on the right side. In fact, what we do get is

$$(p^2 + r^2)z_{42}{}^2 + 2(pq + rs)z_{42}z_{43} + (q^2 + s^2)z_{43}{}^2$$

In particular, therefore, $p^2 + r^2$ must equal 6. We are almost at the last moment of the proof of Theorem 5.9. We just need the following lemma.

Lemma.   The number 6 cannot be written as the sum of two squares of rational numbers.

We shall prove this without reference to any of the theorems in Chapter 2, because we can do so in an elementary fashion. We make an indirect proof, assuming that 6 can be written as a sum of two squares of rational numbers. We continue the notation where we left off above, and write $6 = p^2 + r^2$ with p and r rational. We can set $p = d/f$, $r = e/f$ with d, e, and f integers, by bringing the fractions representing p and r to a common denominator. Then we have

$$(31) \quad 6f^2 = d^2 + e^2$$

Of course, f is not 0.

Among all expressions such as (31), with d, e, and f integers and $f \neq 0$, we may assume that we have selected one with f positive and as small as possible. Now we shall get our contradiction by showing that we can reduce f, after all. We do this by proving that d, e, and f are all divisible by 3. To see this, replace (31) by a congruence mod 3. The left side is congruent to 0. Any square is congruent to 0 or 1 mod 3. For the right side of (31) we thus have the following four possibilities mod 3:

$$0 + 0, 0 + 1, 1 + 0, 1 + 1$$

The last three are ruled out. Hence d and e are both divisible by 3. This makes the right side of (31) divisible by 9, which in turn forces f to be divisible by 3, for otherwise $6f^2$ would be divisible by 3 and not by 9. So d, e, and f are all divisible by 3, as we aimed to prove. This means that we can write $d = 3d_1$, $e = 3e_1$, and $f = 3f_1$ with d_1, e_1, and f_1 integers. If we substitute for d, e, and f in (31) and then cancel 9 throughout, we get

$$6f_1{}^2 = d_1{}^2 + e_1{}^2$$

This is an equation of the same form as (31), but with f replaced by the smaller positive integer f_1. We have the contradiction we wanted, and have thereby completed the proof of the lemma.

With this, the proof of Theorem 5.9 has concluded. In closing, we wish to review the proof of Theorem 5.9 to see what it yields if we start with an affine plane of a general order n.

We label the n^2 points as $x_1, \ldots, x_{n^2}$. There are $n + 1$ directions. For brevity, write $k = n^2 + n + 1$. The directions are assigned the variables from $x_{n^2 + 1}$ to x_k. The $k - 1 = n^2 + n$ lines are augmented by a fictitious line L_k to give k linear expressions $L_1, \ldots, L_k$. Equation (15) becomes

$$L_1{}^2 + \cdots + L_k{}^2 = (n + 1)(x_1{}^2 + \cdots + x_k{}^2) + 2(x_1 x_2 + \cdots + x_{k-1} x_k)$$

The first change in variable, from the x's to the y's, works essentially unchanged, and leads us to

$$(32) \quad L_1{}^2 + \cdots + L_k{}^2 = y_1{}^2 + n(y_2{}^2 + \cdots + y_k{}^2)$$

For the second change of variable we need to generalize (21) and (22), and for this we need to know that any positive integer can be written as a sum of four squares (see Chapter 2, Section 7). Say

$$n = \alpha^2 + \beta^2 + \gamma^2 + \delta^2$$

Then in place of (21) we have

$$\begin{aligned} n(a^2 + b^2 + c^2 + d^2) = {} & (\alpha a - \beta b - \gamma c - \delta d)^2 + (\alpha b + \beta a + \gamma d \\ & - \delta c)^2 + (\alpha c + \gamma a - \beta d + \delta b)^2 + (\alpha d + \delta a \\ & + \beta c - \gamma b)^2 \end{aligned}$$

(The source of this mysterious equation is a number system called the quaternions. It would be fascinating but lengthy to go into the details.)

The second change of variables goes in 4's, so we need information on the values of n and k modulo 4, which we assemble in a table.

Mod 4

n	0	1	2	3
$n^2 + n + 1 = k$	1	3	3	1
$k - 1$	0	2	2	0

In (32) there are $k - 1$ terms (namely, $y_2{}^2, \ldots, y_k{}^2$), which need to be treated in batches of 4, and $k - 1 \equiv 0$ or 2 (mod 4). If $k - 1 \equiv 0$ (mod 4), everything will be used up and we reach no conclusion. If $k - 1 \equiv 2$ (mod 4), there will be two terms left. The upshot:

Suppose there exists an affine plane of order n, where $n \equiv 1$ or 2 (mod 4). Then n is a sum of two squares of rational numbers.

Now it is a fact that if an integer is a sum of two squares of rational numbers then it is a sum of two squares of integers. This is not hard to prove, but we shall not present it. When the statement above is improved in this way, we obtain the splendid theorem of Bruck and Ryser:

Suppose there exists an affine plane of order n, where $n \equiv 1$ or 2 (mod 4). Then n is the sum of two squares of integers.

The first few numbers that cannot be orders of affine planes are: 6, 14, 21, 22, 30, 33, 38, 42, 46, ...

Recall that any power of a prime is the order of an affine plane. For the remaining numbers: 10, 12, 15, 18, 20, 24, 26, 28, 34, 35, 36, 39, 40, 44, 45, 48, 50, . . . no one knows whether an affine plane of that order is possible.

Exercise

Prove that there exist infinitely many integers n such that there is no affine plane of order n. (Hint: Examine $2p$ where p is a prime of the form $4m + 3$. Compare Exercise 4, Section 7, Chapter 2.)

Suggestions for Further Reading

The establishment of modern foundations for Euclidean geometry was the word of the great mathematician David Hilbert. His book on the subject is entitled *Grundlagen der Geometrie* and reached a tenth edition (revised by Bernays) in 1968. An English translation of the tenth edition was published in 1971 under the title *Foundations of Geometry* by the Open Court Co. Parts of the book require considerable mathematical sophistication. The following two treatises on Euclidean geometry give self-contained developments of the subject essentially based on Hilbert's axioms.

K. Borsuk and W. Szmielew, *Foundations of Geometry* (North-Holland, Amsterdam, 1960).

H. G. Forder, *The Foundations of Euclidean Geometry* (Cambridge: Cambridge University Press, 1927).

The following are some references for finite geometry.

W. R. Ballard, *Geometry* (Philadelphia: Saunders, 1970). Chapter 18 deals briskly with affine planes, and then treats projective planes.

A. Tuller, *A Modern Introduction to Geometries* (New York: Van Nostrand Reinhold, 1967). This is a broad survey of geometry which is accessible to a reader with a modest background. There are occasional references to finite geometry.

Anatole Beck et al., *Excursions into Mathematics* (New York: Worth, 1969). Chapter 4 surveys Euclidean and non-Euclidean geometry, and then goes on to finite geometry. Further worthwhile references appear in the bibliography for Chapter 4.

Contributions to Geometry, Slaught Memorial Paper no. 4, *American Mathematical Monthly,* supplement to the August–September issue, 1955. This collection of eight papers, while on the whole rather advanced, gives a picture of some modern research in geometry.

Chapter 6

Game Theory

This chapter is devoted to an introduction to the theory of games. The idea that many familiar games could be subjected to a precise mathematical analysis was first explored by Émile Borel, the French mathematician, in 1921. In 1928, in a remarkable paper published in the German journal *Mathematische Annalen,* John von Neumann created a full scale theory of games. Subsequently, he and Oskar Morgenstern published an expanded version in book form. Since then the theory of games has become a substantial branch of mathematics, with a large literature of its own, and with many mathematicians actively engaged in research in the field.

Émile Borel (1871–1956) dominated French mathematics for a generation. In the study of real variables, complex variables, and probability, his name will be encountered repeatedly. Like many other French mathematicians (e.g., Paul Painlevé, who was the Premier of France twice) he entered public life; he was a member of the Chamber of Deputies from 1924 to 1936, and Minister of the Navy in 1925.

John von Neumann (1903–1957) is considered by many to be the most brilliant mathematician of the twentieth century (Hilbert's career spanned both the nineteenth and twentieth centuries). Born in Hungary, he came to

Princeton University in 1930 and joined the faculty of the newly created Institute for Advanced Study in 1933. Every field he explored (Hilbert space, logic, mathematical physics, game theory), he transformed. At the end of his career, he became known to the general public as a member of the Atomic Energy Commission.

1. Probability

In order to develop the foundations of the theory of games, we need some knowledge of probability. It will suffice to take a naive attitude and to study some examples.

We envisage an "experiment" that can have a certain number (finite) of outcomes. Example: toss a coin; it may fall head or tail. Example: throw a die; it may show any number from 1 to 6. We assume with a minimum of discussion that all the outcomes are equally likely; for example, the coin is legitimate and is equally likely to fall either head or tail, the die is as symmetric as human skill can make it. If there are n outcomes, we say that the probability of each is $1/n$.

What do we mean by this? It is hard to make sense of the statement if there is only one trial. In a single toss the result must be head or tail, and that is that. It is on repeated trials, or in the long run, that we expect heads and tails to even out in some sense. Of course we do *not* mean that on 10 tosses 5 heads and 5 tails are virtually certain. An outcome of 6–4 or 7–3 is not unusual. A result of 10 heads in a row might justify slight surprise, and it might be advisable to examine the coin to see if the two sides are really different.

There is a moderately popular fallacy that after a long string of heads, the next toss is more likely to be tails, in order to even things out. This is not at all true. Granted that the coin is fair, each toss is an independent event; the previous history is totally irrelevant, and the chances are 50–50 all over again. On the other hand, if there is some suspicion that the coin is unfair, a wise man will bet heads rather than tails after there has been a long string of heads.

Suppose however that a fair coin is tossed a million times. It would be utterly fantastic for there to be more than 600,000 or fewer than 400,000 heads. It is in this sense that, in the long run, heads and tails will occur with comparable frequency.

Return again to the experiment with n possible equally likely outcomes. Suppose m of these are singled out as "favorable." Then the probability of a favorable outcome is m/n.

EXAMPLE. If two fair coins are tossed there are 4 equally likely results, which we may record as *HH, HT, TH, TT*. The probability of two heads is ¼. The probability of one head and one tail is ²⁄₄ = ½.

EXAMPLE.   Let two dice be rolled. There are 6 possible results for the first die, 6 for the second, hence 36 altogether. For explicitness we list them all.

11	21	31	41	51	61
12	22	32	42	52	62
13	23	33	43	53	63
14	24	34	44	54	64
15	25	35	45	55	65
16	26	36	46	56	66

It is of special interest to consider the total obtained from the two dice. For instance: Of the 36 possibilities there are 6 (16, 25, 34, 43, 52, 61) that total 7. Hence the probability of rolling a 7 with two dice is $\frac{6}{36} = \frac{1}{6}$. The following table gives all the probabilities of totals that can be rolled with two dice.

Number	Probability
2, 12	$\frac{1}{36}$
3, 11	$\frac{2}{36} = \frac{1}{18}$
4, 10	$\frac{3}{36} = \frac{1}{12}$
5, 9	$\frac{4}{36} = \frac{1}{9}$
6, 8	$\frac{5}{36}$
7	$\frac{6}{36} = \frac{1}{6}$

There is an additional aspect that merits a comment. Consider the three ways of rolling 10: 64, 55, and 46. It is twice as probable that a 10 will be obtained by rolling 6 and 4 as by rolling two 5's. Veterans speak of obtaining two 5's as rolling 10 "the hard way."

We conclude this section with a computation of the probability of winning a game that devotees inelegantly call "craps." In this game the person rolling the dice wins outright if he rolls 7 (probability $\frac{1}{6}$) or 11 (probability $\frac{1}{18}$). He loses outright if he rolls 2, 3, or 12. Because we are computing the probability of winning, we omit these. If he rolls 4, 5, 6, 8, 9, or 10, an auxiliary game begins, in which his objective is to repeat his key number (called his "point") before rolling 7. For instance, if an 8 was first rolled, we have to compute the probability of repeating an 8 before a 7 shows up. Now there are 5 ways of rolling 8 and 6 ways of rolling 7. This accounts for 11 of the 36 possibilities. The simplest way of proceeding is to observe that the remaining 25 cases are irrelevant. So we really have just 11 cases to consider, of which 5 are favorable. This gives us the probability $\frac{5}{11}$.

The discussion is the same for 6 as for 8. For 5 or 9 we get 4 favorable cases out of 10, resulting in a probability of $\frac{2}{5}$, and for 4 or 10 we get 3 out of 9, yielding an answer of $\frac{1}{3}$. We now add the various components to get the total probability of winning.

Initial roll	Probability of this roll	Contribution to probability of winning
7	$\frac{1}{6}$	$\frac{1}{6}$
11	$\frac{1}{18}$	$\frac{1}{18}$
6 or 8	$\frac{5}{36}$	$\frac{5}{36} \times \frac{5}{11}$
5 or 9	$\frac{1}{9}$	$\frac{1}{9} \times \frac{2}{5}$
4 or 10	$\frac{1}{12}$	$\frac{1}{12} \times \frac{1}{3}$

The grand total is

$$\tfrac{1}{6} + \tfrac{1}{18} + 2(\tfrac{5}{36} \times \tfrac{5}{11}) + 2(\tfrac{1}{9} \times \tfrac{2}{5})$$
$$+ 2(\tfrac{1}{12} \times \tfrac{1}{3}) = \tfrac{1}{6} + \tfrac{1}{18} + \tfrac{25}{198} + \tfrac{4}{45} + \tfrac{1}{18} = \tfrac{244}{495}$$

a number just a little less than $\frac{1}{2}$. So the game is a reasonable approximation of an even bet. (That the person rolling the dice has slightly less than half a chance of winning contradicts the beliefs of some fanatical crapshooters. Of course we are assuming fair dice, and a fair roll of the dice, assumptions that are perhaps not always fulfilled in real life.)

Exercises

In Exercises 1–5, find the probability of the event in question.
1. Getting at least two heads when three coins are tossed.
2. Getting an odd number when two dice are rolled.
3. Getting a number $\geqq 9$ when two dice are rolled.
4. Getting a number $\geqq 15$ when three dice are rolled.
5. Getting a ball of each color when two balls are drawn from a jar containing three white balls and two black ones.
6. Four coins are tossed. What is the most likely outcome, and what is its probability?

2. Mathematical Expectation

We content ourselves with a simple-minded description and an example.

Envisage an experiment with n possible outcomes; assume that their probabilities are $p_1, \ldots, p_n$ where $p_1 + \cdots + p_n = 1$. Suppose a man is to receive a sum of a_i if the ith event happens (a_i may be positive, zero, or negative). Then we say that $p_1 a_1 + \cdots + p_n a_n$ is his *mathematical expectation*.

EXAMPLE. On a toss of a fair coin, A gets $2 for heads, loses $1 for tails. A's expectation is $\$(\frac{1}{2}(2) - \frac{1}{2}(1)) = 50$ cents.

What does this mean? Of course on a *single* trial, no 50 cent piece will change hands; either $2 will be won or $1 lost. But after a sufficiently long series of plays, it is reasonable to anticipate an average gain not far from 50 cents. Or, to put it differently, a prudent man would be willing to pay up to 50 cents for the privilege of tossing a coin on the above terms, but he would decline to pay more.

Exercises

In Exercises 1–5, find the mathematical expectation.
1. One die is thrown; for each *n* from 1 to 6, the roller is to get $*n* (i.e., he gets $1 if 1 is rolled, $2 if 2 is rolled, . . . , $6 if 6 is rolled).
2. Three coins are tossed; A get $2 for each head that appears.
3. Two dice are thrown; B is to get $2 if the total is prime, and lose $1 if it is composite.
4. A roulette wheel has numbers 1 to 36, and in addition, numbers 0 and 00 which pay nothing. C bets $1. If the wheel spins to an even positive number he gets his $1 back and wins another dollar; if it spins to an odd number or 0 or 00 he loses his dollar.
5. From four tickets numbered 4, 6, 7, and 9, two will be drawn. D is to get $3 if the numbers are relatively prime and lose $2 otherwise.
6. In Las Vegas casinos the game of craps (described at the end of the preceding section) is modified as follows: If you bet against the person rolling the dice the house does not pay off on 12, this being instead a standoff. What is your mathematical expectation if you bet $1 against the roller of the dice? Compare this with your expectation if you bet with him.

3. Preliminary Remarks on Games

In this section we shall delimit the portion of game theory that will be discussed in the succeeding sections. We do this in a numbered series of remarks.

1. There must be at least two players. In a game of solitaire, questions of skillful play and luck (i.e., probability) may be pertinent. But this has nothing to do with a mathematical theory of the conflict of wills among several players, which is our main objective.
2. On the other hand, if there are three or more players, a complication enters that goes beyond our treatment: the possibility that coalitions may form.
3. So we shall confine ourselves to two-person games. Furthermore,

in the terminology now standard in the subject, our games are *zero-sum*. This means that, if the players are A and B, whatever A wins is paid exactly by B, or if A loses he pays his loss precisely to B. (If a two-person game is nonzero-sum, there is in effect a referee, who pays money or collects some according to the outcome of the game. This introduces some aspects of a three-person game, at least to the extent that A and B might form a coalition against the referee!)

4. A game like chess falls within the scope of the theory, but in an uninteresting way. There is no hidden information, every move being made in the full view of both players. (In bridge or poker, on the other hand, the game derives its excitement from the players' partial ignorance of the opponents' cards.) Common sense tells us that to settle the game of chess once and for all we need only examine all possibilities and choose the best. The game continues to be played only because this project is much too vast to be carried out at present.

5. Our final point is a convenient normalization. Although most games consist of a series of moves, alternating between the two players, we can reduce a player's series of moves to a single move by asking him what he would do under all conceivable circumstances. (Admittedly, in a complex game like bridge, this must at present be considered a theoretical rather than a practical reduction.)

Thus we reach our final setup. We have two players, A and B. Each makes a single decision. A picks an integer from 1 to m, and B picks one from 1 to n; the choice is made simultaneously and secretly. Then they compare. If A has picked i and B has picked j, then B pays A the sum of a_{ij}. Here a_{ij} may be positive, zero, or negative; if a_{ij} is negative, we of course mean that A is to pay B the sum $-a_{ij}$.

The game is completely determined by the mn numbers a_{ij} ($i = 1, \ldots, m$, $j = 1, \ldots, n$). It is natural to assemble the numbers in a rectangular array:

A \ B	1	2	$\cdots$	n
1	a_{11}	a_{12}	$\cdots$	a_{1n}
2	a_{21}	a_{22}	$\cdots$	a_{2n}
$\vdots$			$\cdots$	
m	a_{m1}	a_{m2}	$\cdots$	a_{mn}

Such an array of numbers is called a *matrix*.

4. Examples

Our first example is given by

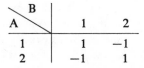

A \ B	1	2
1	1	−1
2	−1	1

A verbal description of the game is as follows: A and B each pick one of the numbers 1, 2; if the numbers agree, A collects a unit from B, while if they disagree, A pays B a unit.

The game is very familiar under the name "matching pennies." We chose not to describe it that way because the suggestion that the players toss a coin at random prejudges the discussion that follows. We wish to emphasize that A and B are *free* to choose 1 or 2 as they please. The game could be played handily by having the players secretly and simultaneously prepare to show one or two fingers.

Concerning one trial of the game there is nothing to say. One dollar will change hands unpredictably, and that is that. It is when the game is played repeatedly that we have something to study. The first fundamental remark is that it is unwise for one player, say A, to keep making the same choice, say 1. Sooner or later B will notice this and start picking exclusively 2, and A will start steadily losing $1 per game.

In Edgar Allan Poe's tale *The Purloined Letter* there is an eloquent description of how the game might work out. The amateur detective in the story is talking to his friend.

> I knew a school-boy about eight years of age, whose success at guessing in the game of 'even and odd' attracted universal admiration. The game is simple and is played with marbles. One player holds in his hand a number of these toys, and demands of another whether that number is even or odd. If the guess is right, the guesser wins one; if wrong, he loses one. The boy to whom I allude won all the marbles of the school. Of course he had some principle of guessing; and this lay in mere observation and admeasurement of the astuteness of his opponents. For example, an arrant simpleton is his opponent, and holding up his closed hand, asks, 'Are they even or odd?' Our school-boy replies, 'odd,' and loses; but upon the second trial he wins, for he then says to himself: 'the simpleton had them even upon the first trial, and his amount of cunning is just sufficient to make him have them odd upon the second; I will therefore guess odd'—he guesses odd, and wins. Now, with a simpleton a degree above the first, he would have reasoned thus: 'This fellow finds that in the first instance I guessed odd, and, in the second, he will propose to himself, upon the first impulse, a simple variation from even to odd, as did the first simpleton; but then a second thought will suggest that this is too simple a variation, and finally he will decide upon putting it even as before. I will therefore guess even';— he guesses even, and wins.

Today we can update Edgar Allan Poe. What the simpleton should do is play 1 and 2, each with probability $\frac{1}{2}$. (He could for instance do this by tossing a coin to decide his move.) Suppose A does this. Then, in the sense of mathematical expectation, it does not matter what B does. If B plays 1, A's expectation is $\frac{1}{2}(1) + \frac{1}{2}(-1) = 0$; if B plays 2, A's expectation is $\frac{1}{2}(-1) + \frac{1}{2}(1) = 0$. Notice that A can announce to B exactly what he is doing without giving B any advantage. The game is a standoff and there is no use even starting to play.

So, the boys in Poe's story who were being outwitted could have staged a recovery merely by pulling out a coin and tossing it to determine their moves.

This example is perhaps too dull to convey much of the flavor of game theory. We see a good deal more in the slightly modified example

A \ B	1	2
1	3	−1
2	−1	1

This is almost the same game as before: A wins when there is a match and loses otherwise. The difference is that when there is a match on 1, A gets a bonus to bring his loot up to $3.

Now a number of challenging questions arise. Presumably A likes this game better than the preceding one (although perhaps even that is not entirely convincing without some proof). But granted this, how much is the modified game worth to him? Or in other words, how much should A prudently pay for the privilege of playing the game? Certainly no more than $3, and presumably a good deal less—$2? $1? 50 cents? 10 cents? Furthermore, how should A play? Should he play 1 more than half the time? Observe that he would like to cash in on his bonus as often as possible. On the other hand, B knows this too, and may accordingly tend to stay away from 1. So should A perhaps anticipate B's reaction and play 1 less than half the time? (Our discussion is beginning to sound like *The Purloined Letter!*) But all this is unconvincing speculation. The marvelous thing is that a precise analysis can yield a precise answer: The game is worth one third of a dollar to A, and he should choose 1 one third of the time.

We prove this by two arguments.

Suppose that A plays 1 with probability $\frac{1}{3}$ and 2 with probability $\frac{2}{3}$. (We shall explain in Section 5 where these numbers come from.) We work out his expectation. If B plays 1, we find $\frac{1}{3}(3) + \frac{2}{3}(-1) = \frac{1}{3}$. If B plays 2 we get $\frac{1}{3}(-1) + \frac{2}{3}(1) =$ the same answer $\frac{1}{3}$. So the outcome is entirely independent of B's course of action. A might as well announce his plans to B and add "It's all settled. Hand over $33\frac{1}{3}$ cents per game."

Can A do even better? We answer this by switching to B's point of view and showing that B can limit his loss per game to $33\frac{1}{3}$ cents. He uses the same probabilities: $\frac{1}{3}$ for 1, and $\frac{2}{3}$ for 2. When A plays 1, his expectation is $\frac{1}{3}(-3) + \frac{2}{3}(1) = -\frac{1}{3}$ (notice the change of sign needed because we are computing on behalf of B). When A plays 2, we get $\frac{1}{3}(1) + \frac{2}{3}(-1) = -\frac{1}{3}$ again.

REMARK. The reader will surely have noticed the fact that the recommended probabilities are exactly the same for B as they are for A. This reflects a special property of this particular game. In the language of matrix theory, the matrix

$$\begin{pmatrix} 3 & -1 \\ -1 & 1 \end{pmatrix}$$

is symmetric in that the same element -1 appears both above and below the diagonal. We shall return to this point in Section 5.

Note again the probabilities that A uses to attain the desired value $\$\frac{1}{3}$: he plays 1 with probability $\frac{1}{3}$ and 2 with probability $\frac{2}{3}$. Furthermore it is true (Exercise 1) that no other choice of probabilities wins him as much as $\$\frac{1}{3}$. So, in his best course of action he plays 1 a good deal less than half the time. We thus get convincing confirmation of our informal guess above that A should tend to stay away from 1.

Exercises

1. In the game

B A	1	2
1	3	-1
2	-1	1

analyzed above, suppose A plays 1 with probability p and 2 with probability $1 - p$, where $0 \leqq p \leqq 1$. Analyze the outcome. (Hint: A's expectation is $4p - 1$ if B plays 1, $1 - 2p$ if B plays 2. These are equal to $\frac{1}{3}$ if $p = \frac{1}{3}$. For $p \neq \frac{1}{3}$, a prudent A will expect to get only the smaller of the two. Find which is smaller, noting that this may depend on whether $p > \frac{1}{3}$ or $p < \frac{1}{3}$. Show that this smaller amount is less than $\frac{1}{3}$.)

2. Find the value to A of the game

B A	1	2
1	k	-1
2	-1	1

where k is a positive number. (Hint: This will be completely analyzed in the next section. But we suggest an informal way of guessing the answer, which can then be fully verified. Try to select the probability p so that A's two expectations $kp - (1 - p)$ and $-p + (1 - p)$ are equal. We guess this common value, which works out to be $(k - 1)/(k + 3)$, to be what A can win. The probabilities p, $1 - p$, used for both A and B, show it to be the true value of the game.)

5. The General Two-by-Two Game

In this section we discuss completely the most general (two-person zero-sum) game given by a two-by-two matrix.

Consider the example

A \ B	1	2
1	3	5
2	1	2

When A examines this game he will reason as follows: "If B happens to play 1, I will do better with 1, getting $3 rather than $1. If B plays 2, I still prefer 1, which yields $5 in place of $2. I will therefore of course always play 1."

B has a similar argument for playing 1, for regardless of what A does he loses less by playing 1.

So the outcome of the game is trivial in that both A and B (if they have minimal intelligence) will always play 1, and A wins $3.

Here is a second example where the analysis is slightly different.

A \ B	1	2
1	3	5
2	2	1

Again A does uniformly better by playing 1, so his choice is obvious. As for B, he prefers losing $3 to losing $5, so if A plays 1 his choice is 1. But if A plays 2, B does better by playing 2, losing $1 rather than $2. So B appears to be torn by conflicting wishes. However, he is capable of appreciating A's point of view. So on a higher level of reasoning he argues that A is always going to pick 1, and does the same himself.

We give a third example where B has a trivial choice, and A, momentarily stumped, also has a trivial choice once he realizes what B is going to do.

A \ B	1	2
1	3	5
2	1	6

In all three cases we are going to say that the game is *silly* (this is not official terminology). If a game is not silly, we call it *sensible*. We proceed to the analysis of sensible games.

Let the four numbers in the two-by-two matrix be a, b, c, and d. Suppose a is the largest of the four, taking account of sign (i.e., a positive number is larger than any negative number). By a harmless renumbering of the moves 1 and 2 of A and/or B, we can arrange to put a in the upper left-hand corner:

A \ B	1	2
1	a	b
2	c	d

Now $a \geq c$. If $b \geq d$, the game is silly from A's point of view. Hence, since we are assuming a sensible game, $b < d$. We can further exclude $a = c$, for the combination $a = c$, $b < d$ means that A would systematically prefer 2 to 1. Thus we have $a > c$, $b < d$ in order for the game to be sensible.

In just the same way, from the assumption that the game is sensible from B's point of view, we get $a > b$, $c < d$. We summarize: Both b and c are smaller than both a and d.

A more or less harmless change can be made in the game. Suppose we replace a, b, c, d by $a - t$, $b - t$, $c - t$, $d - t$, so that the new matrix of the game is

$$\begin{pmatrix} a - t & b - t \\ c - t & d - t \end{pmatrix}$$

Whatever the outcome, the altered game is worth $$t$ less to A.

There is an important number attached to any two-by-two matrix

$$\begin{pmatrix} a & b \\ c & d \end{pmatrix}$$

It is called the *determinant* and it is given by the formula $ad - bc$. The nicest value for a determinant is 0. Our original determinant $ad - bc$ is beyond our control. But if we allow ourselves the subtraction of t mentioned above, perhaps we can alter the determinant to become 0. (Admittedly, this is pulling something of a rabbit out of a hat.) To get the determinant of

$$\begin{pmatrix} a-t & b-t \\ c-t & d-t \end{pmatrix}$$

to be 0, we set down the equation

$$(a-t)(d-t)-(b-t)(c-t)=0$$

or

$$(1) \quad (ad-bc)-t(a+d-b-c)=0$$

Recall that, because the game is sensible, we found that a and d are bigger than b and c. Hence $a+d > b+c$ and it follows that $a+d-b-c$ is positive; in particular it is not 0. Hence we can solve equation (1) for t, getting

$$t = \frac{ad-bc}{a+d-b-c}$$

To simplify notation, let us suppose that the replacement of a, b, c, d by $a-t$, $b-t$, $c-t$, $d-t$ has already been done. In other words, we start all over again with a sensible game

A \ B	1	2
1	a	b
2	c	d

satisfying $ad = bc$. We leave it to the reader (Exercise 1) to prove the following: a and d are positive, and b and c are negative. It will be simpler to follow the rest of the argument if we set $b = -p$, $c = -q$ with p and q positive. The matrix of the game becomes

A \ B	1	2
1	a	$-p$
2	$-q$	d

and we have $ad = pq$. We can now prove that the value of the game is 0 by exhibiting appropriate probabilities for the players to use. A should play 1 and 2 with probabilities $q/(a+q)$ and $a/(a+q)$; then his expectation is 0 no matter what B does, for $qa + a(-q)$ and $q(-p) + ad$ are both 0. B should play 1 and 2 with probabilities $p/(a+p)$ and $a/(a+p)$, and then his expectation is also 0. Notice that A and B use the same probabilities if $p = q$, that is, if the matrix is symmetric, as foreshadowed in the preceding section.

Because we subtracted t from the entries of the original game, we have that the original value to A is given by t. We summarize all this in a theorem, where we have restored a, b, c, d to their original meaning.

Theorem. Let a 2-by-2 sensible game be given by

A \ B	1	2
1	a	b
2	c	d

Then $a + d - b - c \neq 0$ and the value of the game to A is $(ad - bc)/(a + d - b - c)$.

As an illustration of the theorem, consider the game

A \ B	1	2
1	6	2
2	3	11

A brief inspection reveals it to be sensible. The formula $(ad - bc)/(a + d - b - c)$ works out to $(6 \cdot 11 - 2 \cdot 3)/(6 + 11 - 2 - 3) = 60/12 = 5$. The value of the game to A is thus 5.

Let us return to Exercise 2 of the preceding section. The formula yields

$$[k \cdot 1 - (-1)(-1)]/(k + 1 - 1 - 1) = (k - 1)/(k + 3)$$

in agreement with that exercise.

Exercises

1. Let a, b, c, d be real numbers satisfying $ad = bc$. Suppose that a and d are both strictly larger than both b and c. Prove that a and d are positive and that b and c are negative.

2. Which of the following games are silly, and which are sensible? In each case find the value of the game.

(a)

A \ B	1	2
1	2	3
2	4	5

(b)

A \ B	1	2
1	4	3
2	2	5

(c)

A \ B	1	2
1	−2	3
2	4	5

(d)

A \ B	1	2
1	−2	5
2	4	3

3. In the game

A \ B	1	2
1	6	2
2	3	11

(whose value was found above to be 5) find probabilities for A
and B that assure them of this value.

4. In the game given by

B A	1	2
1	5	x
2	4	3

for what values of x is the game sensible?

5. In the game given by

B A	1	2
1	5	x
2	3	4

for what values of x is the game sensible?

6. Find y if the game

B A	1	2
1	1	4
2	y	0

has value 3.

6. Simplified Poker

We shall now discuss a simplified version of poker. The main point
will be to assess the role of bluffing.

It is an empirical observation that most (perhaps all) successful
poker players bluff. So it is interesting to confirm by unimpeachable
mathematics that bluffing is necessary to play the game optimally. If any
reader feels that bluffing is somehow wicked, or at least not sporting, that
feeling should now be dispelled.

Our version of poker is drastically simplified, as compared with the
real thing, but enough survives to give the flavor of the game. There are
two players, A and B. From an enormous stock of cards labeled 1 and 2,
each player is secretly dealt a card. The probability of getting a 1 or a 2
is ½. Player A may now bet "high" or "low."

1. If he bets low, the game is over, A and B compare cards, and
the one with the higher card wins $a (nothing is paid in the event
of a tie).

2. If A bets high, B has two choices. He may give up and (with no
comparison of cards) pay A $a. Or, he may accept the high bet.

Then there is a comparison of cards with the high card winning b, where $b > a$.

Does this game favor A or B? One's instinctive feeling probably is that it should favor A, because he has the privilege of setting the bet high or low. Of course this argument is not too convincing, and in any event it does not indicate how much the game is worth to A, nor how he should play it.

Let us informally discuss A's strategy. Of course his tendency is to bet high with a 2 and low with a 1. Should he be an "honest man" and do this systematically? In that case (as soon as B becomes aware of it) the outcome of the game is cut and dried: B will "call" a high bet if he has a 2 and decline it if he has a 1. We see easily that the expectation of both players is 0.

To do any better, A has to vary his strategy. He has the option of occasionally betting low with a 2, or occasionally betting high with a 1, or both.

Now common sense says that A should never bet low with a 2; after all he can't lose. We shall not speculate on whether this is also true in real poker (when dealt a royal straight flush should you always bet the limit?). At any rate, in our simplified poker, it is easy to prove this (Exercise 1). If preferred, we could bypass this point by making it a game rule that A bets high with a 2.

A similar and even more convincing remark is applicable to B's strategy. If B has a 2, he should always call a high bet.

We come now to the decision of A that really matters: He can, if he wishes, bet high with a 1, despite the fact that he cannot win if the cards are compared. In popular parlance, this is "bluffing."

The other side of the coin is the action B takes when he has a 1 and hears a high bet. He is tempted to surrender and pay a rather than run the risk of a greater loss; but his knowledge that A may be bluffing gives him pause for thought.

We proceed to set up the matrix of the game. We label A's two strategies as H and L (betting high and low). We label B's two strategies as C and T (calling or throwing in). We need to fill in the four blanks in the array below with A's expectation if A and B play as indicated:

A \\ B	C	T
H		
L		

The first case is HC. This means that A bets high if he has a 1 (i.e., he always bets high), and B calls if he has a 1 and hears a high bet. The upshot of this is that the bet is always high and a comparison of cards determines who wins $b. Obviously A's expectation is 0. However, let us set this forth in greater detail, so as to better prepare for the remaining cases. We exhibit a table:

HC	1	1	0
	1	2	$-b$
	2	1	b
	2	2	0

The first row indicates that if A and B both have 1, the payoff is 0 (for A bets high and B calls). In the second row A has 1 and B has 2; A bets high, B of course calls, and A loses $b. In case 21, A bets high, B calls, A collects $b. The final case is an evident tie. The probability of each of the four combinations 1 1, 1 2, 2 1, 2 2 is $\frac{1}{4}$. So A's expectation is $(0 - b + b + 0)/4 = 0$.

We take up case *HT*. The table reads

HT	1	1	a
	1	2	$-b$
	2	1	a
	2	2	0

We explain entries 1 1, 1 2, and 2 1. In case 1 1, A has a 1 but bluffs, B throws his hand in, and A gets $a. In case 1 2, A's bluff is of course called by B who happily holds a 2. A loses $b. In case 2 1, A's high bid is declined by B, who loses $a. We compute A's expectation as $(2a - b)/4$.

We leave it to the reader to satisfy himself that the following tables and the accompanying expectations are correct.

LC	1	1	0
	1	2	$-a$
	2	1	b
	2	2	0

A's expectation: $(b - a)/4$

LT	1	1	0
	1	2	$-a$
	2	1	a
	2	2	0

A's expectation: 0

We now complete the matrix of the game:

A \ B	C	T
H	0	$\dfrac{2a - b}{4}$
L	$\dfrac{b - a}{4}$	0

We recall that $b > a$, so the *LC* entry $(b - a)/4$ is positive. If $b \geq 2a$, so that the *HT* entry is negative or 0, the game is silly from the points of view of both A and B; A will never bluff, and B will never attempt to call a possible bluff. In intuitive terms we might put it this way: If $b \geq 2a$, it is too risky for A to bluff, and likewise it is too risky for B to attempt to call a possible bluff. This sounds quite reasonable, but of course only a mathematical analysis could have told us the precise breaking point.

We continue the discussion under the assumption $2a > b$ (in addition to the running assumption $b > a$). The game matrix thus has the form

$$\begin{pmatrix} 0 & \text{positive} \\ \text{positive} & 0 \end{pmatrix}$$

and it is apparent that the game is sensible.

Our theory in Section 5 supplies the value of the general case. But first let us exhibit a special case. We take $a = 8$, $b = 12$, and find the game matrix to be

A \ B	C	T
H	0	1
L	1	0

The value of the game is $\frac{1}{2}$, and is achieved (by both players) by playing both strategies half the time; that is, A bluffs 50 percent of the time, and B calls bluffs 50 percent of the time.

The formula in Section 5 for the value of the general game yields

$$\frac{0 \cdot 0 - \left(\dfrac{2a - b}{4}\right)\left(\dfrac{b - a}{4}\right)}{0 + 0 - \dfrac{2a - b}{4} - \dfrac{b - a}{4}} = \frac{(2a - b)(b - a)}{4(2a - b + b - a)} = \frac{3ab - 2a^2 - b^2}{4a}$$

So this is the value, from A's point of view, of our simplified version of poker. Note that on putting $a = 8$, $b = 12$ we get

$$\frac{3 \cdot 8 \cdot 12 - 2(8^2) - (12)^2}{4 \cdot 8} = \frac{288 - 128 - 144}{32} = \frac{16}{32} = \frac{1}{2}$$

in agreement with the result obtained above.

Exercises

1. This exercise is devoted to completing the discussion of our game of simplified poker, and to showing meticulously that our assumptions were justified.

We admit the possibility that A may bet low with 2, and that B may fail to call with a 2.

We distinguish all of A's possible choices by the four symbols H_1H_2, H_1L_2, L_1H_2, L_1L_2. Here H_1L_2, for instance, means betting high with 1 and low with 2. For B the choices are C_1C_2, C_1T_2, T_1C_2, T_1T_2.

Complete the table.

A \ B	C_1C_2	C_1T_2	T_1C_2	T_1T_2
H_1H_2				
H_1L_2				
L_1H_2				
L_1L_2				

Show that the first row is better, term-by-term, for A than the second. Likewise the third is termwise better than the fourth. This is a rigorous proof that A should always bet high on 2.

Make a similar discussion for B, as regards the presumed wisdom of his always calling with a 2.

2. Analyze simplified poker with the following change: Both A and B get the 1 card two-thirds of the time and the 2 card one-third of the time; the low and high bets are still a and b. (Hint: The card combinations 11, 12, 21, 22 occur with probabilities $\frac{4}{9}$, $\frac{2}{9}$, $\frac{2}{9}$, $\frac{1}{9}$.)

7. The Game Without a Name

It is beyond the scope of this book to discuss the general (m by n) two-person, zero-sum game. The main theorem, discovered by von Neumann, is that there is a well-defined value for the game, and that suitable strategies for both players demonstrate that the presumed value is indeed the best both players can do.

In this section and the next we present two final examples. Each game treats the two players symmetrically, so that it is "obvious" in advance that the value is 0. Thus our only interest is in finding the strategies the players should use. In both cases the answer seems (to us) unpredictable and surprising.

The first game appears to be nameless. The rules are simple indeed. A and B secretly pick a positive integer; the smaller one wins (equal numbers meaning of course a tie), except that if the smaller number is exactly one smaller, then the player who named it *loses,* and he loses a *double* amount. The matrix of the game is

A \ B	1	2	3	4	5	6	7	...
1	0	−2	1	1	1	1	1	...
2	2	0	−2	1	1	1	1	...
3	−1	2	0	−2	1	1	1	...
4	−1	−1	2	0	−2	1	1	...
5	−1	−1	−1	2	0	−2	1	...
6	−1	−1	−1	−1	2	0	−2	...
7	−1	−1	−1	−1	−1	2	0	...
.	.	.	.	.	.	.	.	...
.	.	.	.	.	.	.	.	...
.	.	.	.	.	.	.	.	...

On the face of it, this is an infinite game (i.e., each player has an infinite number of choices), and should be outside the scope of an account of finite games. However, one of the intriguing aspects is that it turns out to be a finite game.

We present an informal discussion of how we imagine the game would appear to most people (including the authors, before they knew the answer). We quote from an imaginary conversation.

> The game is fundamentally one where low man wins. But of course it will not do to choose 1 all the time; for unless your opponent is a moron he will shortly be picking 2 all the time. However it seems reasonable to pick 1 fairly often; if not half the time, surely at least a quarter of the time, or at any rate as often as any other number.
>
> Probably one should play a lot of 1's, with a judicious admixture of 2's and 3's, and an occasional excursion to larger numbers. How large? There seems to be no convincing reason for drawing the line arbitrarily at some point. Perhaps even the largest of numbers should get played once in a blue moon.

We now announce the answer: The only correct strategy is to play the numbers 1, 2, 3, 4, and 5 with probabilities $\frac{1}{16}$, $\frac{5}{16}$, $\frac{4}{16}$, $\frac{5}{16}$, and $\frac{1}{16}$.

This has the following surprising aspects.

1. The number 1 is played only $\frac{1}{16}$ of the time, whereas both 2 and 4 are played five times as often.
2. 1 and 5 are played equally often.
3. One has to play exactly the numbers 1 to 5 and no higher.
4. The probabilities that must be used are arranged symmetrically about the middle number 3.

We shall not discuss how these probabilities are discovered (but see

Exercise 3). It is, however, not too difficult to check that the probabilities given have the right properties. We exhibit the array

	1	2	3	4	5	6
$\frac{1}{16}$	0	−2	1	1	1	1
$\frac{5}{16}$	2	0	−2	1	1	1
$\frac{4}{16}$	−1	2	0	−2	1	1
$\frac{5}{16}$	−1	−1	2	0	−2	1
$\frac{1}{16}$	−1	−1	−1	2	0	−2
Total	0	0	0	0	0	$1\frac{3}{16}$

The bottom row labeled "Total" represents a computation of A's expectation if he uses the indicated probabilities. We note that if B plays any of 1, . . . , 5, A's expectation is uniformly 0. If B plays 6, A collects a gratifying $\$1\frac{3}{16}$. If B chooses 7 or more, A will do even better (Exercise 1). This indicates that to play satisfactorily B must confine himself to numbers 1 to 5. By symmetry the same is true for A.

Probabilities $\frac{1}{16}$, $\frac{5}{16}$ $\frac{4}{16}$, $\frac{5}{16}$, and $\frac{1}{16}$ are the only ones that guarantee an even break (see Exercise 2).

Exercises

1. Show that if A uses the good probabilities on 1, . . . , 5 and B plays 7 or more, A collects $1 or more for sure.
2. Let A play 1 to 5 with probabilities p_1, p_2, p_3, p_4, p_5, and suppose his expectation is at least 0. Prove that the p's have to be $\frac{1}{16}$, $\frac{5}{16}$, $\frac{4}{16}$, $\frac{5}{16}$, and $\frac{1}{16}$. (Hint: The p's have to satisfy the inequalities

$$
\begin{aligned}
2p_2 - p_3 - p_4 - p_5 &\geqq 0 \\
-2p_1 + 2p_3 - p_4 - p_5 &\geqq 0 \\
p_1 - 2p_2 + 2p_4 - p_5 &\geqq 0 \\
p_1 + p_2 - 2p_3 + 2p_5 &\geqq 0 \\
p_1 + p_2 + p_3 - 2p_4 &\geqq 0
\end{aligned}
$$

If $p_2 = 0$, the first inequality shows that $p_3 = p_4 = p_5 = 0$. This forces $p_1 = 1$, contradicting the second inequality. (Common sense version: If you never play 2, your opponent can thrash you by always playing 1, unless you always play 1, in which case he thrashes you a different way.) Find similar arguments to show all the p's positive. Now common sense says the inequalities must be equalities. For suppose we have $2p_2 - p_3 - p_4 - p_5$ strictly positive. This means B loses money when he plays 1, whereas we know he has a way of breaking even. (Challenge: Make this common sense argument rigorous.) So it remains to make the above inequalities equations and then solve for the p's.)

3. Discuss the nameless game if the double penalty for being exactly one smaller is replaced by a triple penalty. (Hint: As a tentative guess, solve equations:

$$3p_2 - p_3 - p_4 - p_5 = 0$$
$$-3p_1 + 3p_3 - p_4 - p_5 = 0$$
$$p_1 - 3p_2 + 3p_4 - p_5 = 0$$
$$p_1 + p_2 - 3p_3 + 3p_5 = 0$$
$$p_1 + p_2 + p_3 - 3p_4 = 0$$

Then see if the above discussion can be repeated.)

8. Goofspiel

The game of goofspiel is normally played with an ordinary deck of cards. However, the suits are irrelevant; all one needs is a supply of cards numbered from 1 to 13.

Each player begins with 13 cards, bearing the 13 numbers. Another such set of cards is placed in the center. In random order the center cards are put up for bid. For instance, the first center card might be 8. The players now simultaneously and secretly decide what they will bid for it. Perhaps A bids 12 and B bids 7. A wins, and takes possession of the 8 card, which will count 8 points to his credit. (In case of a tie no one wins. With three or more players, we would have to take account of partial ties; however, we are considering only the two-player case.) It is important to observe that A has lost the use of his 12 card for the remainder of the game, and similarly B has lost the use of his 7 card.

The remaining 12 central cards are bid for in the same fashion. (Of course, when the final card is reached the players have no decision to make, their remaining cards are simply compared.) The scores are now totaled. The highest conceivable score is $1 + 2 + \cdots + 13 = 91$, but actually this is impossible. The best one can do is score 90 while the opponent scores 1, which would, incidentally, be quite a tour de force.

A complete analysis of goofspiel would require a formidable computation, perhaps beyond the range of existing computers. All we can do is report the observed behavior of experienced players. Suppose 1 comes up first. It is regarded as extravagant to bid high for it (say as high as 10) and unthinkable to bet 13. One wonders: In the exact analysis of the game is it actually a mistake to bid 13 for 1, or should one do it with some microscopic probability?

On the other hand, if 13 is the first target, successful players have been seen to bid just about anything for it. In particular, they bid 1 for 13 without batting an eyelash, and feel a glow of triumph if they lose to a bid of 13.

So much for true goofspiel. We proceed to a complete analysis of simplified goofspiel, the simplification being that there are 3 cards instead of 13.

Now if the 3 cards are taken to be 1, 2, 3, the game is patently silly: Because 3 is worth as much as 1 and 2 put together, one goes all out by bidding 3 for 3, and then of course 2 for 2 and 1 for 1.

Suppose the central cards are 2, 3, 4 (it is unimportant whether the players' cards are also called 2, 3, 4 but we assume this for simplicity). The game is again silly, although this is no longer obvious at a glance (see Exercise 3, where c, b, a goofspiel is discussed).

Let us study 3, 4, 5 goofspiel, and assume that the first bid is made for 5 (for the other two cases see Exercises 1 and 2). We can look at this as a 3-by-3 game as follows: After the two players have bid for 5, what is left is a subgame where each player holds two cards and bids for 4 and 3 are impending. This subgame is 2-by-2, and its value can be determined by the methods of Section 5, after it is realized that in the subgame the order in which the two cards turn up is irrelevant (why?).

We set up the framework of bids for 5 that needs to be filled in:

A＼B	5	4	3
5			
4			
3			

The diagonal of this 3-by-3 matrix is easy to fill in, for it consists of 0's. For instance: If both A and B bid 5, then neither wins the 5 card, and it remains for them to play a game of 3–4 goofspiel. The value of the latter is plainly 0, and indeed both players will bid 4 for 4 and 3 for 3 unless they suffer an attack of lunacy.

We take up next the case where the bids for 5 are 5 by A and 4 by B. A has won 5 points, and the game that now begins has framework

A＼B	5	3
4		
3		

The intended meaning of the upper-left entry is that A bids 4 and B bids 5 for 4, in which case B wins 4 and there is a tie for the 3. From A's point of view the entry is thus -4. The other three entries are filled in the same way and we find

A＼B	5	3
4	-4	1
3	-1	-3

This 2-by-2 game is sensible. The formula of Section 5 gives

$$\frac{(-4)(-3) - (-1)1}{-4 - 3 - 1 - (-1)} = -\frac{13}{7}$$

for its value. Because A has already won $5 = {}^{35}\!/_{7}$, he shows a net gain of ${}^{22}\!/_{7}$. We enter this in the appropriate spot of the big 3-by-3 matrix, and by symmetry it is clear that $-{}^{22}\!/_{7}$ goes into the diagonally opposite spot. We have reached:

A \ B	5	4	3
5	0	${}^{22}\!/_{7}$	
4	$-{}^{22}\!/_{7}$	0	
3			0

For the two remaining cases we find the subgames to be

A \ B	5	4
4	-7	-3
3	-4	-7

with value $-{}^{37}\!/_{7}$, and

A \ B	5	4
5	-3	1
3	-1	-4

with value $-{}^{13}\!/_{7}$ (just like the first subgame). The big game is now complete.

A \ B	5	4	3
5	0	${}^{22}\!/_{7}$	$-{}^{2}\!/_{7}$
4	$-{}^{22}\!/_{7}$	0	${}^{22}\!/_{7}$
3	${}^{2}\!/_{7}$	$-{}^{22}\!/_{7}$	0

Now our previous experience in Section 7 suggests that we try to find the right probabilities p, q, r by setting all expectations equal to 0. We get the equations

$$-\frac{22}{7} q + \frac{2}{7} r = 0$$

$$\frac{22}{7} p - \frac{22}{7} r = 0$$

$$-\frac{2}{7} p + \frac{22}{7} q = 0$$

In conjunction with $p + q + r = 1$, the solution of these equations is $p = r = \frac{11}{23}$, $q = \frac{1}{23}$.

Summary: If 5 is the first card to be bid for, bid 3 and 5 for it with probability $\frac{11}{23}$ and 4 with probability $\frac{1}{23}$.

Intriguing aspects are:

1. the slightly bizarre denominator 23
2. the (perhaps) unpredictable fact that 3 and 5 get the same probability
3. the fact that 4 must definitely be used, although with a quite small probability

The authors would be curious to see some computations of higher goofspiel. The four-card case would not be terribly exciting, for there is a theorem saying that a sensible game that treats the two players symmetrically must have an odd number of moves. (Remark to those who know matrix theory: Treating the players symmetrically does not mean that the matrix is symmetric, but rather that it is skew-symmetric.) So the next case of genuine interest is five-card goofspiel. Possibly this is too big a job for a hand computation.

Exercises

1. Analyze 3, 4, 5 goofspiel where the first bid is made for 4. (Answer: Bid 5 and 3 with probability $\frac{13}{43}$, 4 with probability $\frac{17}{43}$.)
2. Analyze 3, 4, 5 goofspiel where the first bid is made for 3. (Answer: Bid 5 and 3 with probability $\frac{3}{23}$, 4 with probability $\frac{17}{23}$.)
3. Analyze, c, b, a goofspiel where $a > b > c > 0$ and the first bid is made for a. (Hint: The first subgame to analyze is

B A	a	c
b	$-b$	$b - c$
c	$-(b - c)$	$-c$

This is silly if $b \geq 2c$, otherwise sensible. The value is $-b$ if $b \geq 2c$, $-(b^2 - bc + c^2)/(b + c)$ if $b < 2c$. Next we have

B A	a	b
b	$-b - c$	$-c$
c	$-b$	$-b - c$

which is always sensible, with value $-(b^2 + bc + c^2)/(b + c)$. The third subgame works out the same as the first. So we reach

A \ B	a	b	c
a	0	x	y
b	$-x$	0	x
c	$-y$	$-x$	0

where

$$x = a - b \quad \text{if } b \geqq 2c$$

$$x = a - \frac{b^2 - bc + c^2}{b + c} \quad \text{if } b < 2c$$

$$y = a - \frac{b^2 + bc + c^2}{b + c}$$

Verify that x is always positive. Note that the game is silly if $y \geqq 0$, and that $y > 0$ for $a = 4$, $b = 3$, $c = 2$. When y is negative, say $y = -z$, the correct probabilities to use are

$$\frac{x}{2x + z}, \quad \frac{z}{2x + z}, \quad \frac{x}{2x + z}$$

for bids of a, b, and c.)

Note: The full expression for $x/(2x + z)$ in the case $b < 2c$ works out to be

$$\frac{ab + ac + bc - b^2 - c^2}{ab + ac + 3bc - b^2 - c^2}$$

For $a = 5$, $b = 4$, $c = 3$, this is $2\frac{2}{46}$, checking with our previous $11\frac{1}{23}$. So the denominator 23 is only as "bizarre" as the polynomial $ab + ac + 3bc - b^2 - c^2$.

4. Analyze (c, b, a)-goofspiel when the first bid is made for b or for c.

Suggestions for Further Reading

John von Neumann and Oskar Morgenstern, *Theory of Games and Economic Behavior*, 3rd ed. (Princeton, N.J.: Princeton University Press, 1953). This is the treatise that launched the theory of games as a full-grown discipline. The book is lavish with heuristic motivation, and it indicates extensively how applications to economics are envisaged. Easily readable passages alternate with portions that require considerable mathematical sophistication. Of the 641 pages, 219 are devoted to two-person, zero-sum games. On pages 186–219 there is a study of a version of poker that is much closer to the true game than the simplification presented here.

The article on game theory in the *Encyclopedia Britannica* was written by H. Kuhn and A. W. Tucker, two leading experts in the field. Although somewhat technical, it offers a rewarding synopsis. Their simplified poker allows the second player to raise.

Morton D. Davis, *Game Theory—a Nontechnical Introduction* (New

York: Basic Books, 1970). This book is intended for readers whose mathematical background is slight. The version of poker is the same as ours (with $a = 5$, $b = 8$). Davis also finds *The Purloined Letter* an appropriate prelude. There is an extensive bibliography, including numerous references that emphasize the applications of game theory.

J. D. Williams, *The Compleat Strategyst*, 2nd ed. (New York: McGraw-Hill, 1966). This book is elementary, amusing, and profusely illustrated. An enormous number of two-person, zero-sum games are analyzed.

N. S. Mendelsohn, "A Psychological Game," *American Mathematical Monthly* 53 (1946), 86–88. Mendelsohn, in a short note, discusses the game without a name for the case of an *n*-fold penalty.

Anatole Beck et al., *Excursions into Mathematics* (New York: Worth, 1969). Chapter 5 is an entertaining, elementary ramble through game theory, including the general two-by-two game.

Illuminating historical material concerning the history of game theory appears in the journal *Econometrica* 21 (1953), 95–127.

E. S. Ventsell, *An Introduction to the Theory of Games*, (Lexington, Mass.: D. C. Heath, 1963). This is from a series entitled *Topics of Mathmatics*, translated from the Russian and issued by the Survey of Recent East European Literature at the University of Chicago. Most of this 65-page pamphlet is accessible to a reader familiar only with high school mathematics. The version of poker on pages 24–27 is a little different from ours. At the end there is a brief discussion of infinite games.

Anatol Rapaport, *Two-person Game Theory; The Essential Ideas* (Ann Arbor, Mich.: University of Michigan Press, 1966). A discursive account of two-person games (including nonzero-sum ones) requiring very little mathematics from the reader.

Chapter 7
Infinite Sets

1. Infinite Numbers; Countable Sets

To close the book we return to the topic with which we opened it, the theory of sets. Whereas initially we were interested in some formal, elementary operations on sets and in the notion of function, here we shall be more concerned with something that is perhaps even more basic; namely, the notion of counting. This will not be the first time that counting, or the need for it, has arisen in the book; in fact, in every chapter there has been some point or other where what we have done could be described as counting. The counting that we shall do here, however, will have a twist to it. Instead of contenting ourselves with merely counting out finite things, we will attempt to extend this natural and important idea to counting out sets that are infinite. One could describe what we shall try to do as basically simple—in fact, not only simple but also natural—but despite their essential simplicity, the ideas developed have a rather fine subtlety, which tends to make them quite elusive. There is rather clear evidence for this fact.

First, almost everyone who encounters these ideas for the first time is often at a loss to find something solid and tangible to hold onto. More objectively, although these notions seem so natural today, they were a very long time in coming in the history of mathematics. Relatively speaking, this development in mathematics is not very old, having taken place about 100 years ago. It is due to the insight and genius of the great mathematician, Georg Cantor.

> Georg Ferdinand Ludwig Philipp Cantor was born in St. Petersburg (later called Petrograd, later still called Leningrad) on March 3, 1845. He took his Ph.D. at the University of Berlin in 1867. Two years later he accepted an appointment at the University of Halle, and he remained there until his retirement in 1913. His initial research consisted of a series of fine papers on trigonometric series. This research led him to see the desirability of constructing the theory of infinite numbers, to which this entire chapter is devoted. His ideas engendered vigorous opposition from some mathematicians. But he won stalwart support from others, including Hilbert, who once vowed that no one would ever drive mathematicians out of the paradise created by Cantor. Unfortunately, the last years of his life were made tragic by his mental illness. He died in Halle on January 6, 1918.

In mathematics, as in all human activities, what takes place today is usually an outgrowth of what took place yesterday. Very seldom do we encounter something that represents a true break with what has preceded it, something which—even in retrospect—does not seem a natural consequence of what went on before. Cantor's ideas, even from today's vantage point, represent a true break with precedent, and they still reflect the novelty with which they burst onto the mathematical scene of the early 1870s.

The mark of a good idea in mathematics is found not only in the new results to which it leads but also in what light it sheds on mathematics already known. Cantor's ideas certainly passed this test. Aside from the host of new discoveries it gave rise to, Cantor's way of looking at things gave great new insight into many earlier parts of mathematics. We shall get a glimpse of one such application of the Cantor approach to concrete mathematics when we discuss algebraic and transcendental numbers. On the other hand, some of the questions raised by Cantor's work of 100 years ago have been so difficult that they still remain open. There have been many very great achievements in mathematics in recent years; among these, some of the outstanding achievements can be found in the work of K. Gödel and P. J. Cohen, which deal with the Cantor theory of infinite sets.

Let us begin by asking ourselves a simple question. Suppose we have a primitive man who does not have numbers, or even the notion of numbers, at his disposal. Suppose we give him two piles of things and ask him: Which pile has fewer objects? With what he has available to him, is he in a position to answer this? Clearly yes! What can he do? He can carry out a pairing procedure; that is, he can take one object from one pile, pick up one from the other pile and set them aside. With the new piles he has left he continues the process. With this pairing-off, he can tell which pile has fewer objects; namely, the one that runs out first in this pairing-off.

Now let us analyze what he is doing in terms of the notions and terminology that were introduced in Chapter 1. The two piles of objects are merely two sets S and T. What is the pairing-off procedure? He first picks some object in S, say the element $s \, \varepsilon \, S$, and then pairs it off with some element $t \, \varepsilon \, T$. Let us denote this element t, which is the mate of s in the pairing, by $f(s)$. According to the definition we gave for function in Chapter 1, Section 3, f would be a function from S to T except for the possibility that we may not be able to define f on all of S. When would this happen? It would happen if in the pairing between S and T, we ran out of elements of T first, for then there would be elements of S left for which we could not find mates. In normal words, this would happen when T had fewer elements than S. Suppose, for the sake of argument, that this does not happen. Then f is defined on *all* of S (i.e., for any $s \, \varepsilon \, S$ we have a mate $t = f(s)$ in T), and f is a function from S into T.

What properties does f enjoy? We claim that f is then a one-to-one mapping of S into T; that is, two distinct elements of S have distinct images in T. In fact, this is implicit in our use of "pairing-off"; given an object s in the first pile S, when it has been paired with an object t in the second pile T, they are put aside. Thus t is no longer available for subsequent pairing with any other element of S. Stripped of all these words, what we have is that if $s_1 \neq s_2$ in S, then $f(s_1) \neq f(s_2)$ in T. In the language of Chapter 1, Section 3, this says that f is a one-to-one mapping of S into T. This all took place on the assumption that T ran out first. If S ran out first, we could reverse the roles of S and T and proceed as above, winding up with a one-to-one function from T into S.

Thus we could make the definition that a set S is "smaller than or equal to" another set T if there exists a one-to-one mapping f of S into T. With this kind of formalization of "comparison of size," we could try to measure off *any* set A against *any other* set B. While our discussion of the pairing-off procedure used by the primitive man tacitly assumed that we were dealing with finite sets, the formalization reached after the discussion did not depend on the sets being finite; it could be made for any two sets A and B.

So we start again and define, for two sets S and T, that S is "smaller than or equal to" T, which we write $S \precsim T$, if there exists a one-to-one mapping of S into T. Our intuitive notion of "smaller than or equal to" would demand that certain properties hold for this relation $\precsim$. We list some of these natural properties and try to verify them.

To begin with, we would insist that for any set S, $S \precsim S$; that is, we would want any set S to be "smaller than or equal to" itself. Is this true? All we need is to produce a one-to-one mapping of S into itself. A nice mapping of this kind is always available to us, namely, the *identity mapping* (see Chapter 1, Section 3) $i: S \rightarrow S$ defined by $i(s) = s$ for every $s \, \varepsilon \, S$. Thus indeed $S \precsim S$.

Another property that we would demand is that if $S \precsim T$ and $T \precsim U$, then $S \precsim U$. Is this true in our setting? Since $S \precsim T$, there exists a one-to-one mapping $f: S \rightarrow T$; since $T \precsim U$ there exists a one-to-one mapping $g: T \rightarrow U$. Consider the composite mapping $h: S \rightarrow U$ defined by $h = gf$ for each $s \, \varepsilon \, S$. By the results of Chapter 1, Section 3, h is a one-to-one mapping of S into U. Hence, by the very definition of the relation $\precsim$, we have that $S \precsim U,$ as desired.

There is still a third property of the relation "smaller than or equal to" on which we might insist; namely, if $S \precsim T$ and $T \precsim S,$ then S and T should, in some sense, be "equal." In our situation, what form does this "equality" take on?

First, let us see precisely what the setup is. Since $S \precsim T,$ we have a one-to-one mapping of S into T; since $T \precsim S,$ we have a one-to-one mapping of T into S. So what? Is there anything further we can say here? Indeed we can! This is the content of the Cantor-Bernstein-Schröder theorem* that asserts: In the situation just described, there is a one-to-one mapping of S *onto* T. In the language of Chapter 1, Section 3, S and T are in one-to-one correspondence. (In terms of the pairing-off, this would mean a *perfect* pairing in that we run out of both sets at the same time.) For a finite set, this would say that they have the same number of elements.

We shall not attempt to prove the Cantor-Bernstein-Schröder theorem here (a very nice proof can be found in *"A Survey of Modern Algebra"* by G. Birkhoff and S. MacLane, 3rd ed. (New York: Macmillan, 1965, p. 340). Instead, we use it as the motivation for the reasonable notion of "equality" which we are about to define. More formally:

Definition.   Two sets S and T are said to be *cardinally equivalent* if there exists a *one-to-one* mapping of S onto T.

We denote that the sets S and T are cardinally equivalent by writing $S \sim T$.

Let us return to where all of this originated. To say that finite sets are cardinally equivalent is to say, no more or no less, that they have the same number of elements. It is perfectly reasonable to extend this to the infinite case; that is to say, two infinite sets have the same number of elements when they are cardinally equivalent (i.e., when their elements can be matched against each other). Thus for all sets, cardinal equivalence takes on the significance of "equal number" or "equal size."

*Although Cantor's name appears in the Cantor-Bernstein-Schröder theorem, in point of fact he did not prove it. He wanted very much to prove it, but somehow failed to do so; he then encouraged his student Bernstein to try it. Bernstein came up with an easy proof. Schröder proved it independently.

At this point we can revisit the affine planes of Chapter 5 and view them in a new light. Recall that an affine plane Π is a set in which certain subsets are given special status and called "lines"; the four axioms of Chapter 5 are assumed to hold. Throughout Chapter 5 we assumed Π to be finite. Now we can be more daring and drop this assumption. The statement that any two lines have the same number of points now has meaning for us even when Π is infinite, and the proof given in Theorem 5.1 of Chapter 5, constructing a one-to-one correspondence between the two lines, fits perfectly with the definition above.

A similar case where we can amplify our previous work occurred in connection with Lagrange's theorem (Theorem 4.3, Chapter 4). Let G be any group and H a subgroup of G; we permit G and H to be infinite. For any a and b in G we constructed a one-to-one correspondence between Ha and Hb. This now allows us to assert unreservedly that Ha and Hb have the same number of elements.

Let us now look at two examples coming from the set of integers.

EXAMPLE 1. Let $S = \{1, 2, 3, \ldots\}$ be the set of positive integers and let $T = \{2, 4, 6, \ldots\}$ be the set of even positive integers. Clearly $T \subset S$ and $T \neq S$; thus, in one way, S is strictly larger than T. However, from the point of view of cardinal equivalence, that is, from the point of view of pairing-off, S and T *are* "equal." What we are saying is that $S \sim T$, and that S and T are cardinally equivalent.

To see this we produce a one-to-one mapping f of S *onto* T. Define $f: S \to T$ by $f(n) = 2n$ for any positive integer n. Is f indeed one-to-one and onto? The answer is clearly yes. Because f is a one-to-one correspondence between S and T, S and T are cardinally equivalent.

EXAMPLE 2. Let $S = \{1, 2, 3, \ldots\}$ be the set of positive integers, and let $U = \{0, \pm 1, \pm 2, \pm 3, \ldots\}$ be the set of *all* integers. Here, clearly, $S \subset U$ and $S \neq U$. Yet we still can say that S and U are cardinally equivalent. How? We produce a one-to-one mapping f of S onto U. Pictorially carry out the pairing as follows:

U: 0   1   −1   2   −2   3   −3   ·   ·   ·
      ↓   ↓   ↓   ↓   ↓   ↓   ↓
S: 1   2   3   4   5   6   7   ·   ·   ·

More explicitly, we define $g: U \to S$ by the rules:

$g(0) = 1$
$g(n) = 2n$ if n is a positive integer
$g(-m) = 2m + 1$ if $-m$ is a negative integer (i.e., if m is positive)

Note that $g(75) = 150$, $g(-56) = 113$. We leave it to the reader to verify that the g we have written is indeed a one-to-one mapping of U onto S.

Let us look at Example 2 for just a moment more. What does it really say? It says that we can use the set of positive integers to *enumerate* or to *count off* all the integers. The mechanism for this enumeration is the function g. In our counting off we assign to any integer x the $g(x)$th place. For instance, we would put 10 in the 20th place since $g(10) = 20$, and -50 in the 101st place, since $g(-50) = 2(50) + 1 = 101$.

Here, too, one could say: So what? What is so special about all this? Why shouldn't we be able to use the set of positive integers to count off any set? Well, the surprising (indeed amazing) answer is that there are in fact sets whose elements cannot be enumerated completely, in any way whatsoever, merely by using the set of positive integers. The set of all real numbers—which, after all, is not that bizarre or exotic a set—will be shown to be such a set. This says that we cannot line up the reals in such a way as to count them out: first, second, third, etc., and in doing so run through all the real numbers. Thus, in this new sense of equality—cardinal equivalence—the set of real numbers turns out to be larger than the set of positive integers in that they are *not* cardinally equivalent.

Compare this now with the relation of the set of rational numbers vis-à-vis the set of positive integers. In the sense of cardinal equivalence, we shall see that the set of rationals and the set of positive integers *are* cardinally equivalent. In other words, we *can* line up the rational numbers in some way and say: This element is the first, this one the second, etc., and, in doing so, account for every rational number.

What these things say—in rather imprecise and nonmathematical language—is that there are different levels of infinity, that we can compare "different infinities" as to their "size." In fact, we shall see that given any infinite set, we can use it to construct another infinite set which will be genuinely "larger" than it is in the sense of cardinal equivalence.

Before investigating the things just mentioned, we show that the notion of cardinal equivalence satisfies some requisite properties that we would demand of any notion of "equality."

1. *Every set S is cardinally equivalent to itself; that is, $S \sim S$ for any set S.*

   We do this using the same map, the identity map, as we did in showing that $S \preceq S$ a few pages back.

2. *If $S \sim T$, then $T \sim S$.*

Since $S \sim T$ there exists a one-to-one mapping f of S onto T. But then, as we saw in Chapter 1, Section 3, there exists an inverse mapping $f^{-1}: T \to S$ which is one-to-one and onto. Thus, by the definition of cardinal equivalence, $T \sim S$.

3. *If $S \sim T$ and $T \sim U$, then $S \sim U$.*

Since $S \sim T$, there exists a one-to-one mapping f of S onto T; since $T \sim U$, there exists a one-to-one mapping g of T onto U. By the properties we established for functions in Chapter 1, Section 3, if h is the mapping $h: S \to U$ defined by $h(s) = g(f(s))$ for every $s \varepsilon S$, then h is one-to-one and onto. Hence $S \sim U$.

In much of the preceding general discussion, the set of positive integers $\{1, 2, 3, \ldots\}$ has played a special role, and in what is to follow it will play an equally special role. This is not surprising, because the definition of cardinal equivalence emerged in the attempt to analyze and generalize counting from the finite setting. For this reason we single out the set of positive integers and give it a name, N, which we shall use to denote it.

Those sets that we can enumerate using the elements of N also have a special importance in what is to follow. So we give them their own name.

Definition.   A set is *countable* if it is cardinally equivalent to a subset of N.

Of course the subset of N in the definition is allowed to be all of N. In the two examples presented earlier, we showed that both, in these new terms, are countable. The first example, the set of positive even integers, is already a subset of N, and hence is automatically countable. The second example, the set of all integers, was shown to be in one-to-one correspondence with N, hence it is cardinally equivalent to N and so is countable.

Among the countable sets there are some that are highly particular, the *finite* sets. When do we say that a set is finite? We normally mean by this that we can count out its elements as: first, second, third, . . . , up to some nth one for some positive integer n. This integer n is, in fact, what we call the number of elements in that finite set. Put more succinctly, a set S is finite if for some positive integer n, S can be put into one-to-one correspondence with (i.e., is cardinally equivalent to) the set $\{1, 2, 3, \ldots, n\}$. We also declare the empty set to be finite.

A set, naturally enough, is called an *infinite* set if it is not finite. We introduce one more such name for sets. We call a set *countably infinite* if it is countable but is not finite.

We claim that if S is a countably infinite subset of N, then S is cardinally equivalent to N. Because S is a nonempty set of positive integers, by Property L of Chapter 2, Section 4, S has a smallest element, s_1. If S_1 is the set of elements in S except s_1, S_1 has a smallest element, s_2. Clearly, s_2 is the second smallest element of S. Continuing this way, we have that the elements of S can be ordered as $s_1 < s_2 < s_3 < \cdots < s_n < \cdots$. The mapping $f: s_n \rightarrow n$ is a one-to-one mapping of S onto N. Hence S is cardinally equivalent to N.

If T is a countably infinite set, then T is in one-to-one correspondence with an infinite subset S of N. Since as we have just seen, $S \sim N$, and since $T \sim S$, by the properties developed for cardinal equivalence, $T \sim N$. Thus any countably infinite set is cardinally equivalent to N. Therefore, in speaking about cardinal equivalence, the only infinite subset of N that will be needed and used as a measuring stick will be the set N itself.

If T is a countable set, then there is a one-to-one correspondence $f: T \rightarrow N$, of T into N. If T is countably infinite, then from the discussion above, we may assume that there is a one-to-one mapping f of T *onto* N. Let $S \subset T$ be a subset of the countable set T. Consider the values $f(s)$, $s \varepsilon S$ where f is the one-to-one mapping of T onto N. Clearly, considering f restricted to S, we get a one-to-one mapping of S into N. *Hence S is countable.* If, in addition, S is infinite, S must be countably infinite, and so, cardinally equivalent to N. We summarize this in the statement: *A subset of a countable set is countable. An infinite subset of a countable (hence countably infinite) set is cardinally equivalent to N and thus is countably infinite.*

We introduce the standard description that mathematicians give for the statement that a set S is cardinally equivalent to N. They say that S *has cardinality* $\aleph_0$. ($\aleph$ is the first letter, aleph, of the Hebrew alphabet.)

We now want to show that a familiar set—the set of all rational numbers—is countable (and thus has cardinality $\aleph_0$). We claim that it is enough to show that Q, the set of all positive rationals, is countable. For, if $Q \sim N$, then the set of all negative rationals can be put into one-to-one correspondence with the negative integers. The rational number 0 can be put in correspondence with the integer 0. But then the whole set of rationals can be put into one-to-one correspondence with the set of all integers Z. But, since we already have seen that Z is countable, we know that the set of all rational numbers is countable.

We shall show that Q is countable in two different ways. In each proof we shall use that a positive rational number r can be written as $r = a/b$ where a, b are positive integers with no common factor. We will then say that r is written in *lowest terms*.

Proof 1.   We write out the positive rational numbers according to the following scheme: In the first line we write in increasing order those positive rationals which, when written in lowest terms, have denominator 1 (i.e., the first line shall really be all the positive integers in increasing order). In the second line we write in increasing order those positive rationals which, when written in lowest terms, have denominator 2. In the third line we write in increasing order those positive rationals which, when written in lowest terms, have denominator 3, and so on. Clearly every positive rational number will appear once and only once in this listing. For instance, $r = {}^{131}\!/_{71}$ will appear in the 131st place of the 71st line. However, the number $\frac{4}{6}$ will not appear in the sixth line. Instead, it will appear in the third line, for when we reduce $\frac{4}{6}$ to lowest terms it becomes $\frac{2}{3}$ and so will be in the second place of the third line.

We write it out schematically:

$$
\begin{array}{ccccccc}
1 & 2 & 3 & 4 & 5 & 6 & 7 \quad \cdot \quad \cdot \quad \cdot \\
\tfrac{1}{2} & \tfrac{3}{2} & \tfrac{5}{2} & \tfrac{7}{2} & \tfrac{9}{2} & 1\tfrac{1}{2} & 1\tfrac{3}{2} \quad \cdot \quad \cdot \quad \cdot \\
\tfrac{1}{3} & \tfrac{2}{3} & \tfrac{4}{3} & \tfrac{5}{3} & \tfrac{7}{3} & \tfrac{8}{3} & 1\tfrac{0}{3} \quad \cdot \quad \cdot \quad \cdot \\
\tfrac{1}{4} & \tfrac{3}{4} & \tfrac{5}{4} & \tfrac{7}{4} & \tfrac{9}{4} & 1\tfrac{1}{4} & 1\tfrac{3}{4} \quad \cdot \quad \cdot \quad \cdot \\
\cdot \\
\cdot \\
\cdot
\end{array}
$$

We now number the positive rationals as follows:

$$s_1 = 1, \; s_2 = 2, \; s_3 = \tfrac{1}{2}, \; s_4 = 3, \; s_5 = \tfrac{3}{2}, \; s_6 = \tfrac{1}{3}, \; s_7 = \tfrac{1}{4}, \; s_8 = \tfrac{5}{2} \cdots$$

What are we doing, or better still, how are we counting them off? Again, schematically, we are numbering them using successive diagonals; follow the arrows!

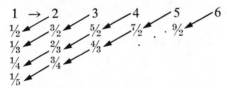

This exhibits a one-to-one correspondence between Q and N. Hence Q is countable.

Proof 2.   In this proof we shall use the fact that a positive integer greater than 1 has a unique factorization into a product of prime powers. Recall that this was proved in Chapter 2, Section 4, Theorem 2.5.

We define the mapping $f: Q \to N$ as follows: If $r \, \varepsilon \, Q$, write $r = a/b$ in lowest terms. Let $f(r) = 2^a 3^b$. This clearly maps Q into N. Why does it do so in a one-to-one fashion? Suppose r and s are distinct rational numbers and $r = a/b$ in lowest terms, $s = c/d$ in lowest terms. Then according to the recipe we gave for f, $f(r) = 2^a 3^b$, $f(s) = 2^c 3^d$. If the integers $f(r)$, $f(s)$ are equal then $2^a 3^b = 2^c 3^d$. By the uniqueness of factorization into prime powers, this would force $a = c$, $b = d$ and so $r = a/b = c/d = s$, contrary to $r \neq s$. Thus f is a one-to-one mapping of Q into N. But then, by the definition of countability, Q is a countable set.

We summarize what we have just done into the

Theorem. The set of all rational numbers is countable (and so has cardinality $\aleph_0$).

We use the spirit of the first proof to prove a related result. Let $A_1, A_2, A_3, \ldots, A_n, \ldots$ be countable sets. Just as we defined the union of a finite number of sets in Chapter 1, Section 1, we can define the union A of $A_1, A_2, \ldots, A_n, \ldots$ as the set of all elements a where a is in at least one A_j. *We shall prove that A is countable.*

We don't want to enter into all the little details of the proof so let us make the following simplifying assumptions about the setup:

1. No two distinct A_i, A_j, have an element in common; that is, $A_i \cap A_j = \phi$, the null set, if $i \neq j$.
2. For $i = 1, 2, \ldots, n, \ldots, A_i$ is a countably infinite set.

Actually, in these assumptions, we are really making A as large as possible, and we will still show that it is countable. If assumption 1 or assumption 2 fails, all it would do would be to *decrease* the total number of elements in A.

Since A_1 is a countably infinite set, we can list its elements as a_{11}, $a_{12}, a_{13}, \ldots, a_{1m}, \ldots$. Likewise, since A_2 is countably infinite, we can list its elements as $a_{21}, a_{22}, a_{23}, \ldots, a_{2m}, \ldots$. More generally, we can list the elements of A_i as $a_{i1}, a_{i2}, \ldots, a_{im}, \ldots$. Now with this notation, we can write out the elements of A, the union of the A_i's as:

$$
\begin{array}{cccccccc}
a_{11} & a_{12} & a_{13} & a_{14} & \cdot\ \cdot\ \cdot & a_{1m} & \cdot\ \cdot\ \cdot \\
a_{21} & a_{22} & a_{23} & a_{24} & \cdot\ \cdot\ \cdot & a_{2m} & \cdot\ \cdot\ \cdot \\
a_{31} & a_{32} & a_{33} & a_{34} & \cdot\ \cdot\ \cdot & a_{3m} & \cdot\ \cdot\ \cdot \\
a_{41} & a_{42} & a_{43} & a_{44} & \cdot\ \cdot\ \cdot & a_{4m} & \cdot\ \cdot\ \cdot \\
\cdot & \cdot & \cdot & \cdot & & \cdot & \\
\cdot & \cdot & \cdot & \cdot & & \cdot & \\
\end{array}
$$

By assumption 1, every element in A appears once and only once in the infinite square array just displayed. Just as in the first proof that we gave for the countability of Q, we list the elements of A as follows:

$$a_{11}, a_{12}, a_{21}, a_{13}, a_{22}, a_{31}, a_{14}, a_{23}, a_{32}, a_{41}, \ldots$$

In this counting out process we are again making use of successive diagonals.

We can actually write out the formula for the function f from A to N being used in the above process. In fact, it is easy to show that $f(a_{ij}) = [(i + j - 1)^2 + i - j + 1]/2$. However, the description of the process is so clear and simple that it seems somewhat pedantic to make use of this explicit formula.

Of course we can also use the idea of the second proof that we gave for the countability of Q, the rational numbers, in the context at hand now. Define $g: A \rightarrow N$ by the rule $g(a_{ij}) = 2^i 3^j$. Here, too, (and for the same reason) g turns out to be a one-to-one mapping of A into N. Thus we see A to be countable also from this point of view. At any rate, we certainly do have that A, which is the union of the countable sets A_1, A_2, . . . , A_n, . . . , is itself a countable set.

What we have shown, namely that the union of the countable sets A_1, A_2, . . . , A_n, . . . is again countable, is usually stated as a theorem.

Theorem. A countable union of countable sets is countable.

Note that from the theorem just proved, it is quite easy to rederive the countability of Q. In Q let A_n be the set of those rationals which, when written in lowest terms, have denominator n. Q is certainly the union of A_1, A_2, . . . , A_n, . . . ; moreover, it is not difficult to show that each A_i is countable. Hence Q itself is countable. Thus by now we have actually proved the countability of Q three times.

There are two special cases of the theorem that have some independent interest.

1. The union of a countable number of finite sets is countable.
2. The union of a finite number of countable sets is countable.

If Z is the set of all integers, then $Z = N \cup M \cup \{0\}$, where M is the set of negative integers. Since N, M, and $\{0\}$ are countable, remark (2) above tells us again that Z is countable. In the same manner we get, knowing the countability of Q, the countability of the set of all rational numbers.

One of the early striking successes of Cantor's work was in its application to algebraic numbers. In this application, a crucial role is played by the theorem that a countable union of countable sets is countable. We shall present this key result about algebraic numbers shortly, but first we digress for a brief discussion of algebraic and transcendental numbers.

A real number a is said to be an *algebraic number* (or merely, algebraic) if we can find, for some integer $n \geq 1$, integers $\alpha_0, \alpha_1, \ldots, \alpha_n$, not all 0, such that $\alpha_0 a^n + \alpha_1 a^{n-1} + \cdots + \alpha_{n-1} a + \alpha_n = 0$. Recall that in Chapter 1, Section 3, we discussed *polynomials*, that is, functions of the form

$$p(x) = \beta_0 x^n + \beta_1 x^{n-1} + \cdots + \beta_{n-1} x + \beta_n$$

The $\beta_0, \beta_1, \ldots, \beta_n$ are called the *coefficients* of $p(x)$ and, when $\beta_0 \neq 0$, n is called the *degree* of $p(x)$. In other words, the degree of $p(x)$ is the highest power of x that really occurs in the expression of $p(x)$. A number, b, is called a root of $p(x)$ if it satisfies $p(x)$ in the sense that

$$p(b) = \beta_0 b^n + \beta_1 b^{n-1} + \cdots + \beta_{n-1} b + \beta_n = 0$$

Thus an algebraic number, by definition, is a real number that is a root of some polynomial, of degree at least 1, having integer coefficients.

In what is to follow we shall need the following result (which we state without proof): *A polynomial of degree n has at most n roots.*

Before going any further, let us see a few algebraic numbers.

1. If r is any rational number, then r is algebraic. For we know that $r = c/d$ where c, d are integers. Hence $dr = c$, and so r is a root of the polynomial $p(x) = dx - c$, of degree 1, with integer coefficients.

2. The number $a = \sqrt{2}$ is algebraic; for a satisfies $a^2 = 2$ and so is a root of the polynomial $p(x) = x^2 - 2$, of degree 2, with integer coefficients.

3. The number $b = \sqrt{1 + \sqrt[3]{5}}$ is algebraic. For, $b^2 = 1 + \sqrt[3]{5}$, hence $b^2 - 1 = \sqrt[3]{5}$. But then, $(b^2 - 1)^3 = 5$. Expanding this, we get $b^6 - 3b^4 + 3b^2 - 6 = 0$. Thus b is a root of the polynomial $q(x) = x^6 - 3x^4 + 3x^2 - 6$, of degree 6, with integer coefficients.

A real number that is not algebraic is called *transcendental*. At the present moment we don't know that there are *any* transcendental numbers. In the next section we shall see that they are plentiful. Perhaps a few words on the history of the affair might be of interest.

In 1844 Liouville (1809–1882) gave a criterion that a number be transcendental. Using his criterion he was able to construct transcendental numbers at will. For instance, he showed that the real number t with decimal expansion

$$t = .10100100000010 \ldots 01 \ldots$$

(here the number of zeros between successive ones goes like 1!, 2!, 3!, . . . , $n!$, . . .) is transcendental. However, little was known about whether some "familiar" numbers were transcendental. The first breakthrough in this direction was made by Hermite (1822–1901) in 1873 when he proved that e, the base of the natural logarithms, was transcendental. In 1882 Lindemann (1852–1939) showed that π was transcendental. A corollary to this result of Lindemann is the fact that by use of straight-edge and compass alone it is impossible to construct a square whose area equals that of a circle of radius 1 (this was the famous old Greek problem of squaring the circle).

The next development came from the great mathematician Hilbert (1862–1943). At a Paris Conference of 1900 he proposed 23 problems that could lead to rich research areas in mathematics. One of these, the seventh, was to show that a^b is transcendental if a, b are both algebraic, with $a \neq 0$, 1 and b irrational. (For instance, a special case would be that $2^{\sqrt{2}}$ is transcendental.) This was proved in 1934, independently, by Gelfond (1906–1968) and Schneider (1911–   ). There are still many open questions about transcendental numbers. For instance, although it is known that e^π is transcendental (this can be deduced from the Gelfond-Schneider theorem by using $e^{i\pi} = -1$) nothing in this regard is known about π^e. As far as is known, π^e may even be rational!

We now return to our discussion of the set of algebraic numbers.

Let M be the set of all polynomials with integer coefficients. We claim that M is countable. Why? We just count out how many polynomials there are of degree n that have integer coefficients. Let M_n be the set of all polynomials of degree n with integer coefficients; a typical element of M_n looks like $q(x) = \alpha_0 x^n + \alpha_1 x^{n-1} + \cdots + \alpha_{n-1}x + \alpha_n$ where the α_i are integers and $\alpha_0 \neq 0$. For each coefficient α_i we can make $\aleph_0$ choices (namely, any integer); from this it follows that we can only choose $\aleph_0$ different sequences $\alpha_0, \alpha_1, \ldots \alpha_n$, and so $\aleph_0$ different polynomials $q(x)$ (see Exercise 5 at the end of this section). Thus M_n is countable. Because $M = M_1 \cup M_2 \cup \cdots \cup M_n \cup \cdots$ is the union of a countable number of countable sets, by the preceding theorem M itself must be countable.

Because M is countable, we can list all its elements in order as $p_1(x), p_2(x), \ldots, p_k(x) \ldots$. Suppose that $p_1(x)$ is of degree m_1, $p_2(x)$ of degree $m_2, \ldots, p_k(x)$ of degree $m_k, \ldots$. By the fact that we asserted earlier, $p_1(x)$ has at most m_1 roots, $p_2(x)$ at most m_2 roots, and so on. If R_1 is the set of roots of $p_1(x)$, R_2 that of $p_2(x), \ldots, R_k$ that of $p_k(x), \ldots$ then, because each R_k is finite, the union $R_1 \cup R_2 \cup \cdots \cup R_k \cup \cdots$ must be countable (again by the preceding theorem). However, from the very definition of algebraic number, this union $R_1 \cup R_2 \cup \cdots \cup R_k \cup \cdots$ comprises all the algebraic numbers. Hence the set of algebraic numbers is countable. For emphasis we state this beautiful and important result as

Theorem. The set of algebraic numbers is countable.

In the next section we shall see that the set of real numbers is not countable. In conjunction with the result just proved, this will show that the set of transcendental numbers is not empty; that is, transcendental numbers must exist. Moreover, it will show that the set of transcendental numbers is not countable. In a certain sense, this says that almost all real numbers are transcendental.

Exercises

1. Show that the following sets are cardinally equivalent to N.
    (a) $S = \{1, 2, 4, 8, 16, \ldots 2^n, \ldots\}$
    (b) $S =$ the set of all rational numbers having odd denominator.
    (c) $S = \{\frac{1}{2}, -\frac{2}{3}, \frac{3}{4}, -\frac{4}{5}, \ldots\}$
    (d) $S =$ the set of all prime numbers.
    (e) $S =$ the set of all prime numbers that can be written as the sum of two squares.

2. Let R be the set of all positive real numbers and let R_1 be the set of all real numbers r where $0 < r < 1$. Show that R and R_1 are cardinally equivalent. (Hint: Examine the function $t/(t + 1)$.)

3. If A, B are sets, let $A \times B =$ the set of pairs (a, b) with $a \ \varepsilon \ A$, $b \ \varepsilon \ B$. Show that if A, B are countable, then $A \times B$ is countable. (Hint: Try to exhibit $A \times B$ as a countable union of countable sets.)

4. Generalize the result of Exercise 3 to show that if $A_1, \ldots, A_n$ are countable, then $A_1 \times \cdots \times A_n$, the set of n-tuples $(a_1, \ldots, a_n)$ where each $a_i \ \varepsilon \ A_i$, is countable.

5. Use Exercise 4 to show that the set of all polynomials of degree n with integer coefficients is a countable set.

6. Let S be the set of all right triangles having integer sides. Show that S is countably infinite.

7. If S is an infinite set, show that there is a subset S_o of S, $S_o \neq S$, such that S and S_o are in one-to-one correspondence.

8. Suppose that we define a set M to be "infinite" if it has a subset $M_o \neq M$ such that M_o and M are in one-to-one correspondence. Show that:
   (a) The set of integers is "infinite."
   (b) If M contains an "infinite" subset, then M is "infinite."
   (c) The union of two "infinite" sets is infinite.
   (Remark: This would probably be taken as the definition of *infinite,* if we were giving a formal treatment of infinite sets.)

9. Show that the following numbers are algebraic:
   (a) $1 - \sqrt{2} + 6\sqrt{3}$
   (b) $\sqrt[3]{7} - \sqrt{11}$
   (c) $\dfrac{1 - \sqrt{3}}{2}$

10. If $a \neq 0$ is algebraic, show that a^{-1} is algebraic.

11. If $a > 0$ is algebraic, show that $\sqrt{a}$ is algebraic.

12. If a is algebraic, show that $1 + a$ is algebraic.

13. If b is transcendental, prove that b^n is transcendental for any positive integer n.

14. If b is transcendental, and $q(x)$ is a polynomial with integral coefficients and positive degree, show that $q(b)$ is transcendental.

2. Uncountable Sets

We begin by summarizing the concepts introduced in the preceding section and the results that were obtained.

The set of positive integers $\{1, 2, 3, \ldots\}$ is denoted by N. A set is called *countable* if it can be placed in a one-to-one correspondence with a subset of N. Two sets A and B are called *cardinally equivalent* if there exists a one-to-one correspondence between all of A and all of B. Thus we can state the matter as follows: A set is countable if it is cardinally equivalent to a subset of N. All finite sets are countable. An infinite countable set is called *countably infinite*. For these sets it is superfluous to bother with proper subsets of N. In other words, countably infinite sets are those that are cardinally equivalent to all of N. All countably infinite sets have the same number of elements, and this number is denoted by $\aleph_0$.

We found that a set may look considerably larger than N and still have the same number of elements as N. In fact, this happened three times in a row, as we formed three successive enlargements of N: to the set of all integers, to the rational numbers, and finally to the algebraic numbers.

At this point might one begin to think that every set is countable, or that the set of real numbers is countable? This in fact happened to Cantor, the founder of set theory. Evidence for this statement is to be found on page 198 of the biography of Cantor which A. A. Fraenkel (an outstanding expert on set theory) published in the 1930 *Jahresbericht der Deutschen Mathematischen Vereinigung*. Fraenkel refers to a lecture by Cantor on September 24, 1897 at Braunschweig; notes on the lecture were taken by Stäckel and were in the possession of Stäckel's wife. At any rate we have in Cantor's own words in the 1882 *Mathematische Annalen* that he succeeded only after considerable difficulty:

> . . . ob sie gleiche Mächtigkeit haben oder nicht, die *aktuelle* Entscheidung darüber gehört aber in den concreten Fällen oft zu den mühsamsten Aufgaben. So ist es mir erst nach vielen fruchtlosen Versuchen vor acht Jahren . . . gelungen zu zeigen, dass das Linearcontinuum nicht gleiche Mächtigkeit mit der natürlichen Zahlenreihe hat.

In a free translation:

> It is often a most difficult task to decide in a concrete case whether or not two given sets are cardinally equivalent. Thus, eight years ago, it was only after many fruitless attempts that I succeeded in proving that the set of positive integers and the set of real numbers are not cardinally equivalent.

We are lucky that we don't have to go through Cantor's struggles. We can relax and let Cantor show us how to do it. Cantor in fact gave two proofs. His first, published in 1874, was based in a rather sophisticated way on order properties of the real numbers. In 1892 he published a second proof; this is the proof that everyone uses today, and the one we proceed to give.

Before we can compare the set of positive integers with the set of real numbers, we need to say precisely what we mean by a real number. Our decision is to look at a real number as being given by its decimal expansion. In other words, a positive real number is named by putting a finite number of digits before the decimal point, and then an unending sequence of digits after the decimal. An example is the decimal expansion of π:

$$\pi = 3.141592653589793\ldots$$

There are two points to be made before we proceed. The first is that it will be a little more convenient not to bother with any digits before the decimal point. This means that we shall really work with a subset of the real numbers: the set (call it T) of all real numbers x satisfying $0 \leqq x < 1$. This change in the problem works in our favor, so to speak. Our project is to prove that the set R of real numbers is not countable. Suppose that instead of this we prove that T is not countable. Recall that in the preceding section we showed that any subset of a countable set is countable. Because T would be known as an uncountable subset of R, by the above observation it would be out of the question for R to be countable. (As an alternative to this discussion we could exploit Exercise 2 of the preceding section.)

The second point we need to raise is not a change of strategy, but rather a candid acknowledgment of a slight difficulty. We have agreed that a positive real number is given by its decimal expansion, and we have put aside real numbers with digits to the left of the decimal (thereby just keeping the set T of real numbers between 0 and 1, including 0 but not 1). Moreover, any decimal expansion is legal and gives rise to a real number. This sounds like a one-to-one correspondence between the set T and the set of all decimals. But there is a catch. The correspondence is imperfect, because under certain circumstances, two different decimals can give rise to the same real number. This happens in the case of a decimal such as

$$.42599999\ldots$$

which ends in an infinite unbroken string of 9's; it and the decimal

$$.42600000\ldots$$

correspond to the same real number.

Fortunately this difficulty is not at all serious. In the proof we are about to give it will be very simple to stay out of trouble. But in the meantime it would be wise to make the following definite agreement: For real numbers of the kind just mentioned we shall not use the version that ends in 9's. Thus in the example above, the official version of the number will be .426000 . . . rather than .425999 . . . .

With these preliminaries out of the way, we state the first main theorem of this section.

Theorem. The set of all real numbers is not countable.

We proceed to the proof. As was remarked above, instead of proving that R is uncountable we shall prove that T is uncountable, where T is the set of all real numbers x with $0 \leqq x < 1$. The proof will be indirect. In other words, we assume that T is countable, and in due course find a contradiction. Because T is of course infinite, to say that T is countable is to say that T admits a one-to-one correspondence with the set of all positive integers. This means that the members of T could be numbered off once and for all, say as indicated by the notation

$$T = \{t_1, t_2, t_3, t_4, \ldots\}$$

As a concrete illustration, the numbers $t_1, t_2, t_3, \ldots$ might, for instance, be

$$t_1 = .271046 \ldots$$
$$t_2 = .335814 \ldots$$
$$t_3 = .890123 \ldots$$
$$t_4 = .666866 \ldots$$
$$t_5 = .151593 \ldots$$
$$\ldots$$

(Of course, this is just an illustration. Because the theorem is true, we cannot possibly exhibit a way of enumerating T!)

Now we come to the beautiful device that Cantor discovered. It is called the "diagonal" method. We prove that the alleged enumeration of T cannot possibly be complete by exhibiting a member of T that is surely missing. We do this by making the new number differ from t_1 in the first decimal place, differ from t_2 in the second decimal place, differ from t_3 in the third decimal place, etc. Call the number we construct this way u. As an explicit way of picking the digits of u we can use the following rule: If the nth digit of t_n is 1, 2, $\ldots$, 9, decrease it by 1; this will be the nth digit of u. If the nth digit of t_n is 0, replace it by 7. (Of course the choice of 7 is fairly arbitrary. The reason we don't use 9 is to make sure we don't get a forbidden decimal that ends in a string of 9's.) In the example given above, the first five digits would be:

$$u = .12778 \ldots$$

The number u cannot possibly be in T, for it is different from every one of the t_i's in some decimal. We have our contradiction, and so we have completed the proof of the theorem.

Now we can return to algebraic and transcendental numbers and view them in a new way. Let R be the set of all real numbers, and A the subset of algebraic numbers. We proved in the preceding section that A is countable. We have just seen that R is uncountable. Of course it follows that A cannot be equal to R. In other words, transcendental numbers exist.

Naturally this is of no help in settling whether or not explicit numbers like e or π or $2^{\sqrt{2}}$ are transcendental. That remains a different kind of question which needs special methods, and the proof is likely to be considerably more difficult than the Cantor diagonal device we have just learned. Still it is charming that in such a simple, conceptual way we can be sure that there exist transcendental numbers.

Sometimes people have gone on to say that Cantor's method is not constructive, and cannot yield an explicit transcendental number. The words of E. T. Bell on page 569 of his *Men of Mathematics* are typical.

> The most remarkable thing about Cantor's proof is that it provides no means whereby a single one of the transcendentals can be constructed.

This is not true. Cantor's idea can be used so as to yield an utterly explicit transcendental number. All we have to do is decide, once and for all, on a definite way of numbering off the set A of algebraic numbers. Because A is countable, we know we can do this. Then we apply the diagonal device to produce, in a unique way, a real number that is different from every member of A. Clearly the number constructed in this way is transcendental. It could be computed to any desired number of decimal places. It is as well determined a number as e or π. (We are greatly indebted to R. M. Robinson for this remark that Cantor's method is really constructive.)

We now return to the notions of *cardinal equivalence* and *cardinal number* that were introduced in the preceding section. Recall that two sets B and C are cardinally equivalent if there exists a one-to-one correspondence between B and C. We may also say then that they have the same cardinal number. All countably infinite sets are cardinally equivalent, and they are awarded the cardinal number $\aleph_0$.

Now we have a second infinite cardinal number. It gets attached to the set R of real numbers, and of course equally well to any set cardinally equivalent to R. A popular notation for the cardinal number of R is the letter c, standing for "continuum" (the set of real numbers is represented by the "continuous" set of points on the real line).

We have two infinite cardinal numbers: $\aleph_0$ and c. Of course we cannot be content with this. We ask: Is there a third? The answer is "yes," and we can tell the reader at once where to look for it: Consider the set of all subsets of the real numbers.

This idea is also due to Cantor. He carried through the argument in its natural setting of maximum generality, by comparing any set B with the set of all subsets of B. We shall call this the *power set* of B and denote it by $P(B)$. Some facts about $P(B)$ when B is finite can be found in Chapter 1, Section 2.

The theorem we wish to prove asserts that B and $P(B)$ are never cardinally equivalent; that is, for no set B does there exist a one-to-one correspondence between B and its power set $P(B)$. Actually, the argument reveals a little more, and the extra information is worth recording: There cannot exist a function mapping B onto $P(B)$, whether or not this function is one-to-one.

Theorem.   Let B be any set and $P(B)$ its power set (the set of all subsets of B). Then there does not exist a mapping of B onto $P(B)$.

To prove this, we shall take the following point of view. We shall assume a function F from B to $P(B)$ given, and we shall exhibit an element of $P(B)$ that is missing from the set of values taken on by F. Note that $F(b)$ is defined for every $b \; \varepsilon \; B$, and that the value $F(b)$ of F at b is a subset of B. With $B = \{1, 2, 3\}$ we give two illustrations of such functions, calling them F_1 and F_2.

$$F_1(1) = \{2, 3\}, \qquad F_2(1) = \{1, 2\}$$
$$F_1(2) = \{3\}, \qquad F_2(2) = \phi$$
$$F_1(3) = \{1, 2, 3\}, \qquad F_2(3) = \{2, 3\}$$

Now we come to Cantor's clever idea. For a given element b in B, it might happen that b lies in the set $F(b)$ or that it does not. Collect all instances of the second kind of element, and write U for this subset of B. In other words, U is the set of all b in B with $b \notin F(b)$. Let us work this out for the two examples above, writing U_1 and U_2 for the sets we get. Since $F_1(1) = \{2, 3\}$ we have that 1 does not lie in $F_1(1)$. Similarly 2 does not lie in $F_1(2) = \{3\}$. But 3 does lie in $F_1(3) = \{1, 2, 3\}$. Thus U_1 has 1 and 2 in it but not 3, so that $U_1 = \{1, 2\}$. Similarly the reader should work out that $U_2 = \{2\}$.

Before continuing the proof we make a side remark that is not essential. The reader might wonder about our having singled out U as the set of all b with $b \notin F(b)$. Why not the set (say V) of all b with $b \; \varepsilon \; F(b)$? With the two examples above, we find $V_1 = \{3\}$ and $V_2 = \{1, 3\}$. Note that V_1 is a value taken on by F_1, since $F_1(2) = \{3\}$, and that V_2 is not a value taken on by F_2, since $F_2(1)$, $F_2(2)$, and $F_2(3)$ are all different from $\{1, 3\}$. Thus we cannot say definitely whether or not V is always taken on by F; this can be true in certain cases and false in others.

But it is a fact that U is *never taken on by F*. Notice that in the examples it checks out that $U_1 = \{1, 2\}$ is not a value of F_1 and $U_2 = \{2\}$ is not one of F_2.

Let us prove that U is not a value taken on by F. If it is, then $U = F(c)$ for some c in B. Now there are two possibilities: Either c is in U or it is not in U. Remember before we proceed that U is the set of all b in B with $b \notin F(b)$; in words, U is the set of all elements of B that are *not members of their corresponding sets*. Now let us try out the hypothesis $c \; \varepsilon \; U$. This means, by the criterion for membership in U, that $c \notin F(c)$. But U *is* $F(c)$. Hence we also have $c \; \varepsilon \; F(c)$. This contradiction drives us to the alternative $c \notin U$. Again, by the criterion for membership in U, this means $c \varepsilon F(c)$. But $U = F(c)$, so $c \notin F(c)$. We again have both $c \varepsilon F(c)$ and $c \notin F(c)$, again a contradiction. We have exhausted all possibilities and have thus shown that no element c in B can exist with $F(c) = U$. In other words, U is not an image under F of any element of B, hence F cannot be onto, and so the theorem is proved.

This is as far as we shall carry our detailed development of the theory of infinite numbers. We shall conclude by indicating informally some of the directions the subject can take from this point on.

Let us recall that in Chapter 1 we used the notation $N(A)$ for the number of elements in a finite set A. Now that we have established a working notion of the number of elements in an infinite set, it is reasonable to extend the notation to infinite sets. Thus for the set Z of integers we have $N(Z) = \aleph_0$, and for the set R of real numbers we have $N(R) = c$.

Let A be any set and $P(A)$ its power set. The theorem we have just proved shows that $N(A) \neq N(P(A))$. We can say more. It is possible to give a precise meaning to the statement that one infinite number is bigger than another, but we shall not enter into the details. When this is done, we can assert that $N(P(A)) > N(A)$.

In Chapter 1, Section 2, we investigated the number of elements in $P(A)$ in the case where A is a finite set. In fact we found the following: If A has n elements, then $P(A)$ has 2^n elements. In other words, $N(A) = n$ implies $N(P(A)) = 2^n$, if n is finite. This suggests that something of the same sort might be equally true when A is infinite.

There is a fast way to make it true for infinite numbers as well as finite numbers—do it by outright definition. (Perhaps this may strike the reader as a useless bit of "cheating" that cannot get us anywhere. But it is a fact that mathematics often makes real progress by arguments that start from a judicious definition that may look artificial.) If $\aleph$ is the cardinal number of an infinite set A, we define $2^\aleph$ to be the cardinal number of the power set $P(A)$ of A. Then we are able to write the statement $2^\aleph > \aleph$ for *any* cardinal number $\aleph$ (finite or infinite). It is always true, although we have not fully discussed its meaning in the infinite case.

But there is more to be said. Let $\aleph$ and $\aleph^*$ be any two cardinal numbers. There is a way of defining $\aleph^{*\aleph}$ so that all the usual properties hold. If, in particular, $\aleph^* = 2$, this definition of $2^\aleph$ agrees with the power set interpretation we gave above.

There is another way of looking at Cantor's theorem that $2^\aleph$ is always bigger than $\aleph$: It says that given any cardinal number we can construct a bigger cardinal number. In other words, there is no largest cardinal number. In studying this theorem, it occurred to Bertrand Russell to try it out on the "universal" set: the set of which absolutely everything is a member.

It would seem that no one could invent a cardinal number larger than the cardinal number of this universal set. Yet Cantor's proof does just that. What goes wrong?

> Bertrand Russell (1872–1970) was a monumental figure whose contributions made a deep imprint on the twentieth century. His early work in logic was climaxed by the publication (with Alfred North Whitehead) of *Principia Mathematica* in 1910–1913 (2nd ed., 1925). His opposition to World War I led to a six-month prison sentence in 1918; there he wrote *Introduction to Mathematical Philosophy*. The following sampling from his books indicates his wide range of interests: *Unpopular Essays* (1950), *The Impact of Science on Society* (1952), *Why I Am Not a Christian* (1957), *War Crimes in Viet Nam* (1967). His autobiography appeared in two volumes in 1967–1968. He was awarded the Nobel prize for literature in 1950. In 1931 he became the third Earl Russell.

Here is what happens when you try out Cantor's proof on the universal set. You are led to consider the possibility of a set being a member of itself, a "self-member" for short. Given a set A, we start listing its members: x, y, z, etc. Perhaps, as we go through the list of members of A, we find A itself! To be sure, this is not something we expect to see every day. But it seems reasonable at least to think of the possibility. In fact, let us call a set that is a self-member "unpleasant," because it does seem a little bizarre. Ordinary sets that are not self-members will be called "pleasant." In other words, a set A is pleasant if $A \notin A$ and unpleasant if $A \varepsilon A$. Now all Russell did was to gather all the pleasant sets, which surely seems a plausible thing to do. The Russell set R is the set of all pleasant sets, that is, the set of all sets A with $A \notin A$.

Now let us try to discover whether or not R is pleasant. We do it in exactly the same way as in the proof of the preceding theorem. Suppose first that R is pleasant. That means $R \notin R$. But remember: The members of R are exactly all pleasant sets, so if R is pleasant, we have on the contrary $R \varepsilon R$. So it must be the case that R is unpleasant. By definition of "unpleasant," this means $R \varepsilon R$; but the members of R are all pleasant, so $R \varepsilon R$ forces R to be pleasant. We are trapped again. The mere act of forming Russell's set R has brought about a contradiction! Thus we have a paradox—it is called *Russell's paradox*.

The same idea can be presented in a simple verbal way. Look at the following two sentences:

The sentence after this is true. The sentence preceding this is false.

Try to figure out whether the first sentence is true or false.

Russell's paradox set in motion a chain reaction of developments in the foundations of mathematics, which remains a subject of intensive study to this day.

There is a further fascinating question that arose in the very earliest stages of Cantor's work. Recall that the first infinite cardinal number we saw was $\aleph_0$, the cardinal number of the set of integers. The second was c, the cardinal number of the set of real numbers. (By the way, it is a fact that $2^{\aleph_0} = c$.) We proved that $c \neq \aleph_0$, and in fact $c > \aleph_0$. Moreover $\aleph_0$ is the smallest infinite cardinal number. The question is this: Is c the next smallest after $\aleph_0$? In other words, does there exist a cardinal number lying strictly between $\aleph_0$ and c? Cantor thought the answer was "no," and he put a good deal of effort into an attempt to prove it. This statement has since been called the "continuum hypothesis." Cantor would perhaps be surprised to learn what happened to the problem, for it is "undecidable." On the basis of the foundations of today's mathematics, it is not possible to prove the continuum hypothesis, and it is not possible to disprove it. This is known as a consequence of the work of Kurt Gödel and Paul Cohen. In 1938 Gödel showed that the continuum hypothesis cannot be disproved, and in 1963 Cohen showed that it cannot be proved.

> Kurt Gödel (1906–   ) made immensely important contributions to mathematical logic, climaxed by his proof in 1938 that the continuum hypothesis is consistent with present-day axioms for mathematics. He is on the faculty of the Institute for Advanced Study in Princeton, N.J.
>
> Paul J. Cohen (1934–   ) took his Ph.D. at the University of Chicago in 1958. After several noteworthy contributions to analysis, his interests shifted to logic, and in 1963 he completed the proof of the independence of the continuum hypothesis. He is on the faculty of Stanford University.

Index

74 75 76 77 9 8 7 6 5 4 3 2 1